THE DESIGN OF COMMUNICATING SYSTEMS

A SYSTEM ENGINEERING APPROACH

THE KLUWER INTERNATIONAL SERIES IN ENGINEERING AND COMPUTER SCIENCE

VLSI, COMPUTER ARCHITECTURE AND DIGITAL SIGNAL PROCESSING
Consulting Editor
Jonathan Allen

Latest Titles

THE DESIGN OF COMMUNICATING SYSTEMS

A SYSTEM ENGINEERING APPROACH

by

C.J. Koomen
Philips Communication Systems
Hilversum, The Netherlands
and
Eindhoven University of Technology
Department of Electrical Engineering

KLUWER ACADEMIC PUBLISHERS
Boston/London/Dordrecht

Distributors for North America:
Kluwer Academic Publishers
101 Philip Drive
Assinippi Park
Norwell, Massachusetts 02061 USA

Distributors for all other countries:
Kluwer Academic Publishers Group
Distribution Centre
Post Office Box 322
3300 AH Dordrecht, THE NETHERLANDS

Library of Congress Cataloging-in-Publication Data

Koomen, C.J. (Cees-Jan), 1947 -
    The design of communicating systems : a system engineering
approach / by C.J. Koomen.
        p. cm. - - (The Kluwer international series in engineering and
computer science. VLSI, computer architecture, and digital signal
processing)
        Includes bibliographical references and index.
        ISBN 0-7923-9203-5
        1. Systems engineering. 2. System design. 3. Telecommunication
systems - - Design and construction. I. Title. II. Series.
TA168.K67 1991
621.382'01'1- -dc20   91-21720
                                                          CIP

Copyright   1991 by Kluwer Academic Publishers

Printed on acid-free paper.

Printed in the United States of America

To my wife Tineke and my sons
who offered me the time
to write this book.

To my parents.

Contents

Preface

"The professional schools will resume their professional responsibilities just to the degree that they can discover a science of design, a body of intellectually tough, partly formalizable, partly empirical teachable doctrine about the design process."
[H.A. Simon, 1968]

Design is aimed at the transformation or translation of a specification or high-level description into a description in terms of some real-world primitives. As such it involves the removal of the uncertainty about the way in which a required system can be realized.

To optimally support the design of systems, we must look at the design process as a whole and at the strong relationship that exists between a designer, the applied design method, the required design tools and the ways in which designs can be expressed. This book focuses on that relationship.

The application field we are concerned with is the design of systems in which the communication between system elements is a major design feature. Examples of such communicating systems are: communication protocols, telephone exchange control systems, process control systems, highly modular systems, embedded software, interactive systems, and VLSI systems. In summary, we are concerned with systems in which *concurrency* plays a major role (concurrency defines the mutual relationship between the activities in the different parts of a system or within a collection of systems).

"Be assured that complete attention to formalism is neither necessary nor desirable. Formality alone is inadequate, because it leads to incomprehensible detail; common sense and intuition alone are inadequate because they allow too many errors and bad designs."
[D. Gries, 1981]

Designing is also an art, a creative act. Within an industrial context, however, the design of digital or computer based systems should involve a purposeful activity; one might speak of *guided* creativity. Hence, we need a scientific basis to design systems and the construction of design methods for those systems, combining formal scientific reasoning with pragmatic views on the design process. Reducing development times and increasing a designer's productivity and the quality of designs are the major targets of such an approach.

The designer establishes the link between a system specification and the final realization, between the *what* and the *how*. In this book we will study this link. The material in this book has been developed over the years at Philips Research Laboratories and is being used for lectures on System Technology at the Eindhoven University of Technology.

Why this book ? And who should read it ?

This book intends to establish a link between theoretical work on formalisms to specify the behaviour of communicating systems, and its application in the fields that were indicated earlier. In this way, a better understanding emerges of the required properties and features of communication formalisms, while at the same time designers who work in the above fields may benefit by applying the results in their own work.

The book is therefore intended for students and designers with an interest in the application of formalisms for designing communicating systems and the methodology for structuring design processes for such systems. Throughout the text exercises are given, and the reader can use these for further training.

The organization of this book,

The material is organized around the *basic design cycle*, which is the central theme of this book and will be introduced in *Chapter 1*.

The first part of this book focuses on *EXPRESSING DESIGNS*. In *Chapter 2* we will introduce *CCS* (Calculus of Communicating Systems). We will use this formalism to express specifications and designs. *Verification* is the subject of *Chapter 3*, while in *Chapter 4* we will *supplement the formalism* with some features that will enable us to deal with certain important properties of systems, such as queues and time behaviour. *Chapter 5* deals with *synthesis*. Verification and synthesis are the two important activities of the basic design cycle. The relationship with other design formalisms, in particular *SDL* and *Petri Nets*, is the subject of *Chapters 6 and 7*.

A DESIGN METHOD FOR COMMUNICATING SYSTEMS is the subject of the second part of this book. *Chapter 8* gives the fundamental principles and outlines the design method, which again is constructed using the *basic design cycle*. Design processes are represented in terms of *meta programs* using a simple language *DPDL* (Design Process Description Language). Then in *Chapters 9, 10 and 11* the design of the call processing part of a telephone exchange is treated as an example, using a the material from all previous chapters. The example is rather extensive and requires some determination on the part of the reader. The material combines methodology and formalism and shows their application in a non-trivial case.

Finally, in the third part, we focus on *DESIGN, CREATIVITY AND LEARN-ING*. It contains a methodological treatment of design processes, aspects of creativity, computability, and the use of computers in the design process. In *Chapter 12* we study the occurrence of iterations. In principle, such iterations are due to learning taking place during the design process. We will consider these iterations from the point of view of Godel's results, computability, chaos, and entropy. We will present some information laws that govern design processes. We will discuss some limits for the mechanization of design processes. In *Chapter 13* the results of the previous chapter are used to reason about CAD and learning. In *Chapter 14* we will give a metric for expressing the information content of designs. It can be used to calculate the information generated by a designer and to tune design iterations.

Acknowledgement

Writing this book would not have been possible without the cooperation and support of many people. First of all, while I was at Philips Research Laboratories, Michel Sintzoff suggested me to look into CCS. This book shows that I followed that advice. I also like to thank Robin Milner for his comments on parts of the original manuscript and for the discussions we have had in the past.

Two of my graduate students, Peter de Graaff and Robert Huis in 't Veld, have been inspiring during many discussions. In particular I like to thank Robert for his large support and active help in getting the book ready. Another graduate student, Willem Rovers, contributed in the final editing. Johan Wevers supported me in the lay-out phase.

My son Casper-Jan was a great help to me by preparing all the drawings using his skills on the PC.

I like to thank the students of the 1990/1991 course on System Technology for helping me find and correct the errors in the original version. I trust this effort has stimulated them to read the text with even greater concentration. And I thank all the others who directly or indirectly have been helpful in preparing the material.

Part I

EXPRESSING DESIGNS

Chapter 1

The Basic Design Cycle

1.1 The two directions of the design process

Design is a creative activity. Simon [Sim68] considers the design process as
a kind of search process involving the selection of an acceptable solution (he
speaks of a "satisfying" solution) from a set of potential solutions which have
been considered by the designer.

During this creative search process, involving the creation and valuation of possible alternatives, the designer is constantly improving design knowledge. In terms
of Simon's search paradigm this means that the designer is developing criteria
with which he or she is able to obtain alternatives and select the most acceptable
(satisfying) solution. Design can therefore be considered as a process involving
two basic activities (see *Figure 1.1*):

(a) the creation of possible solutions

This involves *construction, refinement,* or *synthesis*. The designer uses knowledge
as well as available building blocks to do this.

(b) selecting an acceptable solution.

This involves the selection of the solution which satisfies the design specification,
or which maximizes a certain utility function. This activity requires the *evaluation* or *verification* of the result in terms of the specification. Synthesis (how
to get what I want?) and verification (is what I have got the correct thing?) go
hand in hand and are the two basic activities carried out by the designer.

One could object that the best way is not to check (verify) a design after it has
been made, but to synthesize systems such that the correctness of the result
is implicit. We call this approach *correctness by construction*. However, this
type of synthesis is not always possible; it depends on the availability of correct
construction rules (one should realize that such rules or transformations have
to be verified too). Hence, correctness by construction and verification are very

much related. These two basic activities yield the following model of a **basic design cycle**:

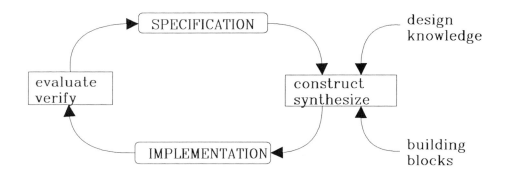

Figure 1.1: The basic design cycle.

Consider the following example. Suppose a designer has to design a blue cube. The designer knows about geometric objects in general, that a cube has certain properties, and that it can be cut out of a piece of wood. The designer may therefore decompose the specification into:

S = S1 ∧ S2

where

S1 : object = cube
S2 : colour of object = blue.

Next, the designer specifies two construction methods $M1$ and $M2$. Method $M1$ involves the cutting of the cube from a piece of wood, while $M2$ involves painting the cube blue. The effect of $M1$ and $M2$ can be stated in the following form:

$$M1 \quad \frac{\text{a piece of wood}}{\text{a wooden cube}} \qquad M2 \quad \frac{\text{object}}{\text{blue object}}$$

where the expression above the line defines the situation that should exist before the method can be applied; the expression below the line describes the situation that exists after the method has been applied. We observe that $M1$ implements $S1$, whereas $M2$ implements $S2$. Hence, this is an example of correctness by construction.

However, suppose we would like to show correctness separately. We could build an object recognizer, i.e. a pattern recognizing machine $V1$, which would have stored in its memory descriptions of objects such as 'cube'. We would need a second piece of equipment $V2$ which would perform a spectral analysis; based on its findings it would produce a statement about the colour of an object presented to it. Hence, if we have designed a blue wooden cube, then machine $V1$ would produce the output statement 'cube', whereas $V2$ would produce the output statement 'blue'. These outputs can be directly linked to the specifications $S1$ and $S2$ and by deduction we conclude that the object that has been created indeed satisfies the specification S.

Design involves the creation of the necessary detailed information about the way a required system can be realized in terms of known techniques, rules, processes and building blocks. Hence, the creation of possible solutions implies that the designer creates or adds such detailed information during the execution of design cycles. In *Section 1.4* we will briefly discuss the underlying mechanisms.

1.2 Writing specifications and implementations

A **specification** is a prescription of a desired system. A specification is either expressed informally (e.g. in natural language), or is given as a *formal specification*, in which case specifications are defined as mathematical objects [Win90].

An **implementation** (or *design*) is a description of a coherent set of subsystems, each given in terms of a specification, such that these subsystems together satisfy the specification. As long as an implementation does not satisfy its specification, one should in fact speak of a *potential implementation*. For reasons of notational simplicity we shall speak of *implementation* also in that case. One can distinguish two major ways to write specifications and implementations:

(i) techniques based on the definition of properties of systems
    (these are *prescriptive* techniques)

For instance: if we want to define an error-free communication channel in terms of properties, we state that if a message is offered to such communication channel, then this message will eventually be available at the output of the channel.

(ii) techniques based on abstract models
    (which *describe* a system's behaviour)

In this case a system specification represents an abstract model of the system. For example, a finite-state machine can be used to describe the behaviour of a communication channel in terms of its state changes, depending on whether messages have been received or sent.

Specifications based on prescriptive techniques do not address implementation details, whereas specifications based on abstract models already suggest an implementation mechanism (such as the finite state machine). The technique that we will use in this book is of the abstract model type since it resembles the finite-state machine approach. The technique will be introduced in *Chapter 2*.

1.3   Verifications which shape design processes

Verification is the activity which checks the consistency between a specification
and a potential implementation. We speak of a *correct design step* if the verification shows that the implementation is indeed consistent with, or satisfies, the
specification. If both the specification and the implementation are represented as
mathematical objects, then we can use *formal verification* techniques; formally
verifying a design requires the construction of a mathematical proof.

Verification closes the design cycle. If the resulting implementation is correct,
then the designer will proceed to another design cycle. If not, the current design
cycle will be repeated using the results and gained knowledge from the previous
iteration. Repeating the current design cycle may require changing its specification. In that case, the effect of the verification is migrated to other parts of the
system or to other phases in the design process.

A **design process** is a collection of basic design cycles. During a design process,
design cycles are created and, if verification is successful, they may persist over
time. Other design cycles may become obsolete due to the effect of verifications
elsewhere and the subsequent updating of the collection of design cycles. Hence,
the actual design process is a dynamic collection of design cycles, reflecting a
process of learning by the designer who tries and evaluates possible solutions.
We will consider a more extended model of the design process in *Chapter 8*.

The properties of a system can be specified along a number of dimensions, such
as: functional behaviour, performance, reliability, dissipation, etc. This means
that verification should take place along each of these dimensions. Verification
of the correctness of the functional behaviour can be done in various ways:

1. by means of **inspection**

    Inspection is an informal technique in which the designer inspects the de-
    sign or program and tries to reason that the design is correct. This tech-
    nique is acceptable for small designs, but practice has shown that even in
    small designs, design errors may remain using this method. *Design walk-
    throughs* are based on inspection carried out by a team of designers.

2. by means of **simulation**

    For larger designs, such as integrated circuits or printed circuit boards, one
    usually applies simulation to show that the design satisfies its specification.
    Often, no formal specification is available; only a document stating the
    required functionality (in a natural language form) may be available. In
    that case simulation is used as a tool for verification as well as to validate
    the specification.

3. by means of **formal verification techniques**

Here, design objects are mathematical objects. This renders the possibility to apply mathematical proof techniques for verification.

Both inspection and simulation are widely used techniques. The use of formal verification techniques as well as the use of transformations [DF84] are not yet everyday's practice. However, the use of high-level hardware description languages such as ELLA [Pra86] or VHDL [LSU90] in combination with automated synthesis tools looks promising, although this approach does not yet imply mathematical proof techniques. In this book, we will focus on one possible formal technique, where verification will be based on equivalent behaviours of systems. We distinguish two types of formal verification:

(i) techniques based on **the definition of properties of systems**

As we saw earlier, specifications prescribe properties of a system. These properties are given as axioms. Verification means constructing proofs to show that each specification axiom is a theorem derivable from the implementation axioms [Gor83].

(ii) techniques based on **abstract models**

Formal verification means that, given two abstract models (one for the specification and another, less abstract, model for the implementation), one has to show that the implementation satisfies the specification in the sense that the two are equivalent models of the required system. This requires a mapping of the implementation back onto the specification using certain transformation rules. Another way of putting this is that verification aims to find an abstraction of the implementation such that this abstraction is equivalent to the specification. In *Chapter 3,* we will return to this view on verification.

1.4   Mechanisms for synthesis and verification

As we have discussed earlier, one can discern two main directions of information flow throughout the design process. One direction involves the creation of possible solutions by adding detail or refining the specifications. The other direction involves the removal of information for verification purposes.

A number of mechanisms by which synthesis or verification take place can be distinguished. In order to do this we need to have a very simple model of a specification. Since a specification is put in terms of natural language or in terms of some mathematical notation, we can simply state that a specification is a set of statements. Using this simplified view of what constitutes a specification, one can discern the following basic mechanisms in the direction towards greater detail:

1. **refinement**: statements are added to a set of statements.

2. **decomposition**: a set of statements is split into a number of subsets.

3. **synthesis**: this involves both decomposition and refinement.

And in the other direction we find the complementary mechanisms:

4. **simplification**: statements are removed from a set of statements

5. **composition**: sets of statements are combined into a single set.

6. **reduction**: sets of statements are replaced by an equivalent set of less detailed statements.

Complementary mechanisms are: refinement and simplification, decomposition and composition, synthesis and reduction. We will be applying these mechanisms throughout this book; additional references to decomposition are [Par72] and [PS75]

1.5   Trends in specification techniques

In the early days of computers, programs were hard-coded as bits in memory. This process became cumbersome with increasing size of the programs. For that reason, machine instructions were written in a short-hand notation using mnemonics, which evolved into assembly languages. Still later, higher-level programming languages were introduced. Such programming languages underwent an evolution in terms of their structuring and modularization capabilities. However, they all have in common that a programmer has to indicate very precisely how the solution is to be executed; no details can be left out, although for the nature of the solution some details may be irrelevant.

Specification languages allow a designer to express *what* the system should do (leaving out irrelevant details), rather than *how* the system will be implemented. A special class of such specification languages are the *requirements specification languages* [DHL+86, DH87, Zav82], which, in their language constructs, are aimed at supporting the requirements phase preceding system design. *Object oriented design* [CY90, Mrd90] is a specification and design approach which uses abstract models of the objects which a system has to process or handle. Other references to formal and structured approaches to software and system development can be found in [Boe88, Jac86, Jon86, Law90, LB85, LZ77, SP90, MM90, WM85]. An original approach to the role of languages in design processes is presented in [HitV90].

Table 1.1 summarizes the trends in software and hardware specifications. The question may arise as to whether a software specification language coincides

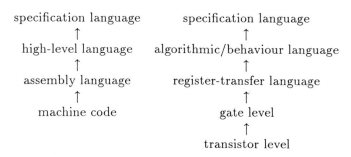

SOFTWARE	HARDWARE
specification language	specification language
↑	↑
high-level language	algorithmic/behaviour language
↑	↑
assembly language	register-transfer language
↑	↑
machine code	gate level
	↑
	transistor level

Table 1.1: Evolution in specifications.

with a similar language for hardware. After all, a program finally runs by the exchange of signals between hardware elements. So we might as well translate a program directly into hardware; this is the idea underlying the concept of *silicon compilation* [dMCG+90] and logic synthesis [Tre87]. A particular function can be realized in terms of various hardware/software combinations. The optimal balance between the hardware and software is determined by cost, performance, compatibility of products, etc.

In the software design area there is a trend from sequential programming, with an emphasis on techniques for data structures and data manipulations, towards parallel or concurrent programming techniques. In those techniques programs are often described in terms of communicating processes.

In the hardware design area, a similar trend can be observed. The need for high-level specification techniques emerged not only to master complexity and reduce development times, but also to describe the inherent parallelism in hardware. This has led to techniques for describing hardware in terms of communicating processes. By describing systems in terms of their communicating subsystems or processes, *the software and hardware design areas meet.* Examples of such approaches can be found in [Mil89] and [Hoa85].

There is a second reason why techniques to specify systems in terms of their communications are of interest. When designing complex systems, decomposing a system in terms of interconnected subsystems is a well accepted principle. This, in turn, triggers a need for using a formalism that enables one to express relationships between subsystems in terms of their communications.

By using a formalism that enables us to express communication, we are in principle able to specify both hardware and software systems. Such formalism also allows us to use a design method based on a decompositional approach, where systems are decomposed into communicating subsystems. In the next chapter we will consider CCS, the Calculus of Communicating Systems. We will use CCS for various reasons:

(i) it is among the first calculi that explicitly deal with communication;

(ii) other calculi are derived from it or have been influenced by the work on CCS;

(iii) using a calculus allows us to manipulate expressions using certain rewrite rules. This allows for the use of mechanical (i.e. computer) support during the specification and design process;

(iv) the calculational approach is very suited to specify abstract models of the type of systems we are interested in;

(v) after learning to apply CCS, the reader may find it relatively easy to adopt other formalisms similar to CCS.

Chapter 2

Calculus of Communicating Systems

2.1 Introduction to CCS

The Calculus of Communicating Systems (CCS) originates from Robin Milner. CCS deals with communication behaviours of (non-deterministic) finite state machines. Among the first publications on CCS is [Mil78]. In 1980, Milner published his first book on CCS [Mil80]; [Mil89] is an update of the previous book with many new results. The calculus is meant as a formal mathematical framework which can be used to specify communicating systems and to verify properties of them. A system is considered to consist of parts, each of which has an own identity, which persists through time. A system will be characterized by the actions it can perform.

CCS is among the first theories dealing in a formal way with communication behaviour of systems. Other approaches which fall into this class are [Hen88], trace theory [Rem85, Kal86], and ACP (Algebra of Communicating Processes) [BK83, Bae86]. Hoare [Hoa85] developed theoretical CSP (Communicating Sequential Processes). [Win90, CP88] contain further references to formal techniques.

In CCS, the objects of behaviour are called **agents**. One can think of an agent as a locus of activity, a process, a computational unit. An agent's behaviour is defined in terms of the actions it can perform. CCS provides mechanisms to represent behaviours as algebraic expressions, to calculate the combined behaviour of communicating agents, to determine whether two agents have equivalent behaviours, and to reduce equations to a minimal form. Although proofs in CCS can be lengthy, the steps are relatively simple and can be mechanized.

We will briefly review CCS. Since our main focus will be on applying CCS, we will not go in great mathematical detail. The interested reader is referred to [Mil89] for a detailed treatment of the calculus. We will consider the objects for

11

which the algebra is defined and the operations defined on these. We will also have a look at several laws which can be used to transform behaviour expressions of agents. These laws play an important role during the formal verification of implementations against specifications as described in e.g. [Koo85a]. In CCS, proofs basically consist of the application of laws which are used as rewrite rules.

CCS is founded on two central ideas. The first stems from the question: "how to describe the behaviour of systems, consisting of subsystems which communicate concurrently?". Milner chose to answer this question by stating that we can describe a concurrent system fully enough by considering the behaviour as experienced by an external observer. Hence, the first idea is that of *observation*. This also allows us to speak of the equivalence of two systems; two systems are indistinguishable if we cannot tell them apart without pulling them apart. *Observation equivalence* is based on this notion of observation.

Although we aim to describe concurrent systems in terms of externally observable behaviour, we also wish to be able to speak about system structure in terms of communicating (sub)systems. Hence, *communication* is the other central idea of CCS. To calculate behaviours, we need an operator to perform *parallel composition*; this operator composes processes, allowing them to communicate.

These two ideas, observation and communication, are really one. For it is supposed that the only way to observe a system is to communicate with it, which makes the observer and system together a larger system. In other words: to place two components in communication (i.e. to compose them) is just to let them observe each other.

2.2 Actions

We first introduce an infinite set $\mathcal{A}$ of **names**; a, b, c,... range over $\mathcal{A}$. We denote by $\bar{\mathcal{A}}$ the set of **co-names**; $\bar{a}, \bar{b}, \bar{c}, \ldots$ will range over $\bar{\mathcal{A}}$. Furthermore, we introduce the set of **labels** $\mathcal{L}$, where $\mathcal{L} = \mathcal{A} \cup \bar{\mathcal{A}}$ .In terms of communication behaviour, labels denote ports via which communication takes place; hence, a communication action via a port labelled x ($x \in \mathcal{L}$) is represented by x. Two complementary labelled ports x and $\bar{x}$ are said to *communicate*. We may interpret x as the **input port** and $\bar{x}$ as the **output port**; however, we will only distinguish between input ports and output ports when such distinction seems necessary.

Suppose we have two agents A and B; agent A has a port labelled x, while agent B has a port labelled $\bar{x}$. Since both actions have complementary labels, by definition they communicate. If such a communication occurs (i.e. A performs x, while B performs $\bar{x}$), we say that the combination of A and B performs a silent or **internal action**. We will use the symbol τ to denote such action. Silent actions play an important role, and their use and meaning will become clear when we start specifying and verifying systems.

We have introduced actions, the set $\mathcal{L}$, and the silent action τ. We now define the set of actions $\mathcal{ACT} = \mathcal{L} \cup \{\tau\}$, to consist of the previously defined labels and the silent action. Furthermore, we need a relabelling function to rename labels in order to connect ports via which we wish agents to communicate. Hence, we define a relabelling function f to be a function from $\mathcal{L}$ to $\mathcal{L}$, such that $f(\bar{a}) = \overline{f(a)}$ for $a \in \mathcal{L}$. The set of labels $\mathcal{L}(A)$ associated with an agent A is called the **sort** of that agent (*Figure 2.1*).

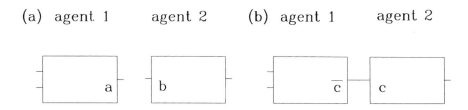

Figure 2.1: *(a)* Two agents before relabelling, and *(b)* after relabelling with the relabelling function f, such that $f(a) = \bar{c}$ and $f(b) = c$

2.3   Representation of behaviours of agents

A **behaviour equation** is an algebraic expression denoting the behaviour of an agent. Such an equation is of the form <agent>=<expression>, where <agent> is an **agent identifier**, and <expression> is an **agent expression**. The symbol "=" should be read as "is defined as". Later, we will introduce another interpretation based on equivalences. We will use capital letters $A, B, C, \ldots$ as agent identifiers. We define $\mathcal{E}$, the set of agent expressions by the following syntax (E, E_1, E_2 are already in $\mathcal{E}$):

$\alpha : E$	Action Prefix ($\alpha \in \mathcal{ACT}$)
$E_1 + E_2$	Summation
$E_1 \mid E_2$	Composition
$E \backslash L$	Restriction ($L \subseteq \mathcal{L}$)
$E[f]$	Relabelling (f a relabelling function)

Both summation and composition are defined here as binary operators. However, for our type of applications (describing systems in terms of communicating subsystems, such as communication protocols) we need + and | to be associative and commutative (in which case it is immaterial in what order subsystems are put together). The reader is referred to [Mil89]. Hence, expressions like $E_1 + E_2 + E_3$ in the case of summation, or $E_1 \mid E_2 \mid E_3$ in he case of composition, are valid expressions.

Equations may be bracketed to indicate the required order for the application of operators. The default order is (in terms of decreasing binding power): restriction and relabelling (strongest binding), action prefix, composition, summation. Thus, for instance,   A + a: B| b: C\ L means A + ((a:B) | (b:(C\ L))) .

From the rule for <expression> we learn that a form $A = A + B$ is possible. We will assume that behaviour equations are guarded [Mil89] (p.65) or guardedly well-defined [Mil80] (p.72). This means that if the same agent identifier occurs at both the left-hand side and the right-hand side of a behaviour equation, then the agent identifier at right-hand side has to be preceded by at least one action; hence, an expression of the form $A = a : A + B$ is guarded.

Exercise 2.1. Put in brackets; a : b : A + τ : B| C\{a} + b: C.   ∎

Behaviour equations can be interpreted as state transition equations in which the left-hand side denotes the current state and the right-hand side gives the state transitions and the next state(s).

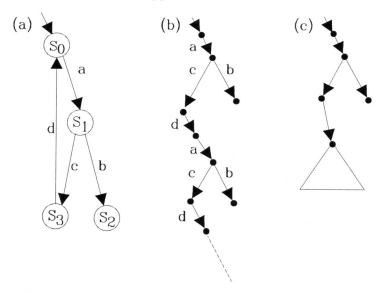

Figure 2.2: (a) a state graph; (b) unfolding of the state graph; (c) a recursive representation.

Consider the state graph of *Figure 2.2(a)*; $S0, S1, S2$ and $S3$ are the states; a, b, c and d denote actions, which are associated with the state transitions. 'Unfolding' the state graph yields the (infinite) tree of *Figure 2.2(b)*. As a short-hand, *Figure 2.2(c)* shows a recursive definition of this tree. The trees of *Figure 2.2(b)* and *Figure 2.2(c)* will be referred to as **behaviour trees**. A behaviour tree represents the behaviour of an agent in terms of all possible sequences of actions that can occur (these sequences correspond to *traces* in [Kal86].)

The behaviour equation of the agent $S0$ of *Figure 2.2* is:

$S0 = a : (b : NIL + c : d : S0)$

NIL is the behaviour with no actions. Alternatively, we may write:

$S0 = a : S1$
$S1 = b : S2 + c : S3$
$S2 = NIL$
$S3 = d : S0$

A behaviour equation is said to be in **normal form** whenever <expression> is a sum of **simple agent expressions**; a simple agent expression consists of an action, followed by (i.e. sequentially composed with) an agent identifier. Hence, the above form is the normal form of the behaviour equation of $S0$.

Exercise 2.2. Put $A = a : (b : c : A + \tau : A)$ in normal form.   ∎

From this example we observe that the phrase 'agent $S0$ will produce agent $S1$ when action a occurs' is equivalent to saying that 'the system will move from state $S0$ to state $S1$ if state transition a occurs'. In the sequel both *agent* and *state* will always be understood to mean an agent in some state.

In summary, behaviour equations will have the following syntax:

<behaviour equation>	⇒	<agent>=<expression>
<expression>	⇒	<s-exp>
	⇒	<expression>|<expression>
	⇒	<expression> + <expression>
	⇒	<expression> \ <labels>
	⇒	<agent>
	⇒	<action>:<expression>
<s-exp>	⇒	<action>:<agent>
<labels>	=	is a set of labels that will be applied for restricting the agent expression.
<agent>	=	agent-identifier

We will use the convention that expressions like $A = \alpha : \beta : \gamma :NIL$ can be abbreviated as $A = \alpha : \beta : \gamma$, i.e. we omit the NIL symbol and assume expressions to end at NIL in that case.

2.4   Semantics of CCS operators

The semantics of CCS is defined by partitioning the set of all agent expressions
into disjoint sets; expressions which belong to the same set are indistinguishable
or, *equivalent*. This partitioning is obtained by defining an equivalence relation
(*observation equivalence*). This equivalence relation is expressed in terms of inference rules which define the operational behaviour of CCS expressions. Inference
rules are of the form:

$$\frac{\text{hypothesis}}{\text{conclusion}}$$

In the hypothesis part the behaviour of the component agents is given; the conclusion part then defines the behaviour of the composite agent. Since we declared
that agents are to be identified with states, we interpret actions as state transitions; the property that an agent in state E_1 can perform an action α and then
proceed to state E_2 will be written as

$$E_1 \xrightarrow{\alpha} E_2 \quad \text{(read: ``E_1 produces E_2 under action α'')}$$
$$\text{(or: ``E_2 derives from E_1 under α'')}$$

For each of the operators (action prefix, summation, composition, restriction
and relabelling) we will give the corresponding inference rules. In each of the
following rules, E, E_1, E_2 and F are agent expressions from $\mathcal{E}$. In the examples,
as shown in the figures, $A, B, \ldots, H$ are particular agents represented by their
behaviour trees. The resulting behaviour tree in one example is used in the next
example.

ACTION PREFIX

Act $\dfrac{}{\alpha : B \xrightarrow{\alpha} B}$ (This rule is in fact an axiom)

Rule **Act** states that an agent of the form $A = \alpha : B$ can execute action α and
then show behaviour B.

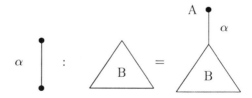

Figure 2.3: Illustration of action-prefix.

The **SUMMATION** $A + B$ combines the behaviours of agents A and B to form
a new agent whose behaviour is either A or B.

$$\textbf{Sum}_1 \quad \frac{E_1 \xrightarrow{\alpha} E_1'}{E_1 + E_2 \xrightarrow{\alpha} E_1'} \qquad \text{and} \qquad \textbf{Sum}_2 \quad \frac{E_2 \xrightarrow{\alpha} E_2'}{E_1 + E_2 \xrightarrow{\alpha} E_2'}$$

Rule $\textbf{Sum}_1$ states that if E_1 produces E_1' under action α, then $E_1 + E_2$ also produces E_1' under the same action. Likewise for $\textbf{Sum}_2$.

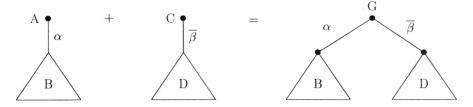

Figure 2.4: Illustration of summation.

The **COMPOSITION** $A \mid B$ of two agents A and B yields another agent, the behaviour of which is the concurrent behaviour of A and B (rules $\textbf{Com}_1$ and $\textbf{Com}_2$), unless communication between A and B occurs (which follows from the complementarity of labels as in $\textbf{Com}_3$).

$$\textbf{Com}_1 \quad \frac{E \xrightarrow{\alpha} E'}{E \mid F \xrightarrow{\alpha} E' \mid F} \qquad \textbf{Com}_2 \quad \frac{F \xrightarrow{\alpha} F'}{E \mid F \xrightarrow{\alpha} E \mid F'}$$

$$\textbf{Com}_3 \quad \frac{E \xrightarrow{l} E', F \xrightarrow{\bar{l}} F'}{E \mid F \xrightarrow{\tau} E' \mid F'} \quad (l \neq \tau)$$

Rules $\textbf{Com}_1$ and $\textbf{Com}_2$ state that the parallel composition of two agents can execute all actions of these agents. In case the agents have complementary actions then in addition to the individual actions ($\textbf{Com}_1$ and $\textbf{Com}_2$), rule $\textbf{Com}_3$ states that in that situation the parallel composition *also* yields a τ action.

Notice that communication between two agents ($\textbf{Com}_3$) is a synchronized action between those agents. Also notice that $\textbf{Com}_3$ involves abstraction since the information about the reason for the occurrence of the τ (namely, the communication via l and $\bar{l}$) is lost. Hence, communication and abstraction are intertwined in this rule. We can separate the two issues by defining a rule $\textbf{Com}_3'$:

$$\textbf{Com}_3' \frac{E \xrightarrow{l} E', F \xrightarrow{\bar{l}} F'}{E \mid F \xrightarrow{(l,\bar{l})} E' \mid F'} \quad (l \neq \tau)$$

which writes the combined action $(l, \bar{l})$ instead of τ; now the abstraction can be done separately using an abstraction function $\textbf{ABS}$, such that $\textbf{ABS}(l, \bar{l}) = \tau$.

RESTRICTION $(A \mid B) \backslash \{b, \bar{b}\}$ means that all individual actions b and $\bar{b}$ are removed from $(A \mid B)$.

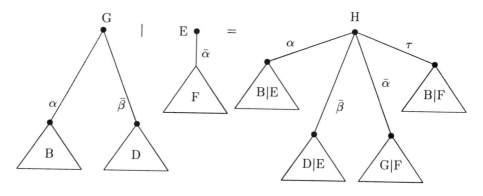

Figure 2.5: Illustration of composition.

Res $\dfrac{E \xrightarrow{\alpha} E'}{E \backslash L \xrightarrow{\alpha} E' \backslash L}$ $(\alpha, \bar{\alpha} \notin L \text{ and } L \subseteq \mathcal{L})$

Rule **Res** states that in the case of our example of *Figure 2.5*, restriction over α means removing all edges labelled with α from the behaviour tree; this is illustrated in *Figure 2.6*. Instead of restrictions like $A \backslash \{\alpha, \bar{\alpha}\}$ it will suffice to write $A \backslash \{\alpha\}$ since we will assume that restriction includes all complementary actions of the label set L in expression $A \backslash L$.

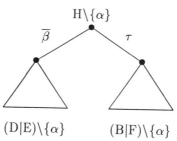

Figure 2.6: Illustration of restriction.

Restriction allows one to abstract from internal details of a system. Composition and restriction are both examples of abstractions; this will prove to be very useful when we want to formally verify correctness. (the relationship between verification and abstraction was already hinted at in *Section 1.3*).

RELABELLING is needed in order to connect ports together by assigning complementary names to these ports. Each relabelling function f should satisfy $f(\bar{a}) = \overline{f(a)}$. The syntax of a relabelling function f, applied to an agent with agent expression E is $E[\alpha/\beta, \gamma/\delta, \ldots]$, which tells us that everywhere in the agent expression E, label β is to be replaced by label α, label δ by label γ, etc.

Rel
$$\frac{E \xrightarrow{\alpha} E'}{E[f] \xrightarrow{f(\alpha)} E'[f]}$$

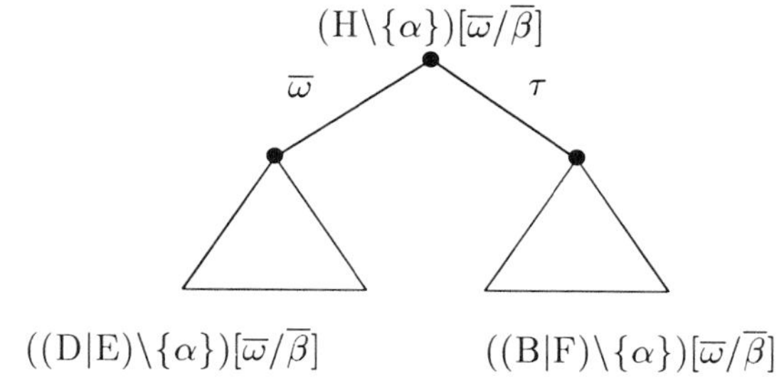

Figure 2.7: Illustration of relabelling.

2.5 Parallel composition of agents

In order to calculate the combined behaviour of two or more communicating agents, Milner developed the **expansion law** [Mil89] (chapter 3.3) and [Mil80] (p.31 and p.82). With the expansion law the behaviours of two or more communicating agents can be combined to form one behaviour, i.e. they can be considered as a single system. The law describes how expressions of the form $D = (A \mid B \mid C)\backslash L$ are to be calculated.

In this new system D, the internal communication between the original systems A, B, and C is defined by L; L is the set of labels corresponding to the ports connecting A, B, and C. The effect of the restriction will be that we abstract from the internal communication between the subsystems A, B, and C; we are only interested in the *effect* of this internal communication at the interface between D and its environment. For example, in the system of *Figure 2.8*, we wish to calculate the behaviour of A, B and C at the interface ports δ and ϵ; we are not interested in the details of the communication at α, β and γ; we are only interested in the effect of these communications at δ and ϵ. We write the behaviour we are interested in as:

$$(A|B|C)\backslash\{\alpha, \beta, \gamma\}$$

indicating that we wish to abstract from the communication at ports α, β and γ. This abstraction will yield τ's at those places in the behaviour equation for the composite system, where communication between A, B and C takes place.

It is important to notice that concurrency within CCS is modeled by taking all possible *interleavings of actions*. This is a consequence of the previously introduced rules, in particular the **Com** rules. We can further illustrate this

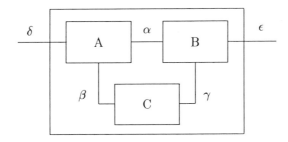

Figure 2.8: Parallel composition of the agents A, B, and C.

as follows. We assume a fine grain time scale upon which we map the possible sequences of actions of the individual subsystems. No two actions will map onto the same point in time unless communication between two subsystems takes place. Hence, we do not model such things as simultaneous actions, unless two subsystems communicate. For example, if we have two agents, of which the behaviours are:

$A = \alpha$ and $B = \gamma$

then (since the agents do not communicate) their combined behaviour, calculated by interleaving their actions is:

$A|B = \alpha : \gamma + \gamma : \alpha$

Whenever two subsystems communicate, we write a τ to denote this communication. For example, if we have two subsystems

$A = \alpha$ and $B = \bar{\alpha}$

then, by taking all *possible* interleavings of their actions yields:

$A|B = \alpha : \bar{\alpha} + \bar{\alpha} : \alpha + \tau$

Hence, given the behaviour $(A \mid B)$, agents A or B either do not communicate (yielding the interleaving of the actions α and $\bar{\alpha}$), or they do communicate (yielding the τ for their synchronized communication). In case we had restriction over the ports α and $\bar{\alpha}$, then expansion would have yielded

$(A|B)\backslash\{\alpha\} = \tau$

since the individual actions are not permitted due to the restriction; however, their communication would be allowed (this is a property of restriction).

Exercise 2.3. Show the above result by using rule **Res**.   ■

2.6 Expansion law

In order to calculate expressions of the form $(A \mid B \mid C) \backslash L$, Milner developed the following expansion law. Let $P = (P_1 \mid \ldots \mid P_n) \backslash L$, with $n \geq 1$. Then

$$P = \sum \{\alpha : (P_1 \mid \ldots \mid P_i' \mid \ldots \mid P_n) \backslash L\} \qquad (P_i \xrightarrow{\alpha} P_i', \alpha \notin (L \cup \bar{L}))$$

$$+ \sum \{\tau : (P_1 \mid \ldots \mid P_i' \mid \ldots \mid P_j' \mid \ldots \mid P_n) \backslash L\}$$

$$(P_i \xrightarrow{\lambda} P_i', P_j \xrightarrow{\bar{\lambda}} P_j', i < j, \lambda \in L)$$

The two parts of the right-hand side represent (i) the actions of the individual subsystems, and (ii) their mutual communication respectively.

In general, we need to perform relabelling in order to connect ports together before we can apply expansion. The general form of the expansion law is as follows. Let $P = (P_1[f_1] \mid \ldots \mid P_n[f_n]) \backslash L$, with $n \geq 1$ and $[f_i]$ a relabelling of $P_i (1 \leq i \leq n)$. Then

$$P = \sum \{f_i(\alpha) : (P_1[f_1] \mid \ldots \mid P_i'[f_i] \mid \ldots \mid P_n[f_n]) \backslash L\}$$

$$(P_i \xrightarrow{\alpha} P_i', f_i(\alpha) \notin (L \cup \bar{L}))$$

$$+ \sum \{\tau : (P_1[f_1] \mid \ldots \mid P_i'[f_i] \mid \ldots \mid P_j'[f_j] \mid \ldots \mid P_n[f_n]) \backslash L\}$$

$$(P_i \xrightarrow{\lambda} P_i', P_j \xrightarrow{\bar{\lambda}} P_j', \ i < j, \lambda \in L, \ f_i(\lambda) = \overline{f_j(\bar{\lambda})})$$

We can state the expansion law in algorithmic terms as follows. First, we define the state of a system as an n-tuple, where each term is the state (the agent identifier) of one of its subsystems. For example, $S = s_1 s_2 \ldots s_n$, where $s_1 s_2 \ldots s_n$ are the states of the n subsystems. Given the behaviour equations of the subsystems in normal form, then the steps are:

step (a) define an initial state for the combined system. For instance, if we wish to calculate $E = (A \mid B \mid C \mid D)$, then $S = ABCD$ initially;

step (b) we apply the first part of the expansion law by listing all possible individual actions in each of the subsystems which are permitted by the restriction. We also write the resulting state of the combined system for each of these actions. Hence, we look for transitions of the type $S \xrightarrow{\alpha} S'$, where $S = s_1 \ldots s_i \ldots s_n$ and $S' = s_1 \ldots s_i' \ldots s_n$. These transitions do not yet involve communication between subsystems.

step (c) we then apply the second part of the expansion law by adding all actions which involve communication between the subsystems, which we denote by silent actions τ. For each of these communications we also write down the resulting state of the combined system.

step (d) repeat steps (b) and (c) for each of the resulting states which have not yet been evaluated in this way.

In the following example, we use an alternative naming scheme for actions where we make it explicit whether a port acts as an input port or as an output port. The action corresponding to a communication via an input port is denoted by the name of the input port, followed by a question mark (like *a.in1?*). Likewise, an output action is written as the name of the corresponding output port, followed by an exclamation mark (like *a.out!*). This naming scheme is convenient wherever the designer wants to indicate explicitly which ports are input and which ports are output. In our example, the connections are given by the block diagram. (Alternatively we could express this by means of a connection table, e.g. *b.out2 = a.in2*, etc.).

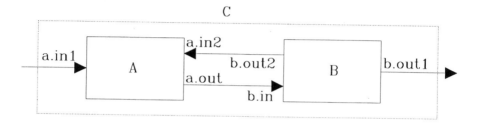

Figure 2.9: A handshake system.

The specifications for A and B are given as follows:

A = a.in1?: a.out!: a.in2?: A
B = b.in?: b.out1!: b.out2!: B

In normal form:

A   = a.in1? : A1   B   = b.in? : B1
A1   = a.out! : A2   B1   = b.out1! : B2
A2   = a.in2? : A   B2   = b.out2! : B

The connection table for A and B is:

a.out = b.in
a.in2 = b.out2

In order to calculate the behaviour of the total system C, we have to evaluate the following expression:

$$C = (A\,[\bar{\alpha}/\text{a.out!}\,,\,\beta/\text{a.in2?}]\ |B\,[\alpha/\text{b.in?},\,\bar{\beta}/\text{b.out2!}])\ \backslash\{\alpha,\beta\}$$

Using the steps as indicated earlier and applying these to the normal forms of A and B yields:

step (a) AB is the initial state;

step (b) we consider those actions which each of the subsystems A and B may wish to execute: this yields actions *a.in1?* and *b.in?* respectively. However, we conclude that action *b.in?* is not permitted by the restriction. The resulting state of C is $A1B$. We write the following behaviour equation:

$$AB = \text{a.in1? : A1B};$$

step (c) At this point no communication can take place; we repeat steps (b) and (c) for state $A1B$.

step (b) From state $A1B$ neither of the two possible actions *a.out!* and *b.in?* can occur individually since they are not permitted by the restriction. We proceed to step (c) for this state.

step (c) Communication can indeed take place and we write down:

$$A1B = \tau : A2B1$$

Repeated application of steps (b) and (c) to the resulting states of system C yields:

$$A2B1 = \text{b.out1! : A2B2}$$
$$A2B2 = \tau : AB$$

Starting with the behaviour equations of A and B in normal form, the last step finishes the expansion yielding the behaviour of C in normal form. Observe that in our example each iteration through steps (b) and (c) yields one action and the resulting global state of system C. In general, we will have a list of actions (and their resulting global states) after each iteration.

Of course we might also have used the original behaviour equations of A and B, and rewrite the expression for C. In that case, the expansion would yield:

$$
\begin{aligned}
C \;=\;& (A|\,B)\backslash\{\alpha,\beta\} \\
=\;& ((\text{a.in1?}{:}\bar{\alpha} : \beta{:}A)|\,(\alpha{:}\text{b.out1!}{:}\bar{\beta}{:}B))\backslash\{\alpha,\beta\} && \text{(substituting the} \\
&&& \text{expressions for } A \text{ and } B) \\
=\;& \text{a.in1?}{:}((\bar{\alpha} : \beta{:}A)|\,(\alpha{:}\text{b.out1!}{:}\bar{\beta}{:}B))\backslash\{\alpha,\beta\}) && \text{(action a.in1?)} \\
=\;& \text{a.in1?}{:}\tau{:}(((\beta{:}A)|(\text{b.out1!}{:}\bar{\beta}{:}B))\backslash\{\alpha,\beta\}) && \text{(communication via } \alpha \\
&&& \text{and } \bar{\alpha}) \\
=\;& \text{a.in1?}{:}\tau{:}\text{b.out1!}{:}(((\beta{:}A)|\,(\bar{\beta}{:}B))\backslash\{\alpha,\beta\}) && \text{(action b.out1!)} \\
=\;& \text{a.in1?}{:}\tau{:}\text{b.out1!}{:}\tau{:}((A|\,B)\backslash\{\alpha,\beta\}) && \text{(communication via } \beta \\
&&& \text{and } \bar{\beta}) \\
=\;& \text{a.in1?}{:}\tau{:}\text{b.out1!}{:}\tau{:}C && \text{(substituting } C)
\end{aligned}
$$

Apart from the silent actions τ, the composite system C shows the behaviour *a.in1? : b.out1!* repeatedly as would be expected of a handshaking mechanism.

It is straightforward to interpret the steps of the expansion in terms of constructing the state graph of the composite system. *Figure 2.10* shows how the state

graph of our handshake system C can be constructed on the basis of the steps
of the expansion.

Exercise 2.4. Given A=$\alpha + \beta$,B=$\bar{\alpha} + \gamma$,C=$\bar{\beta}$. Show that the composite system
$C=(A| B| C)\backslash\{\alpha\}$ can be expanded into:

$$C=\beta : (\gamma : \bar{\beta}+\bar{\beta} : \gamma)+\gamma : (\beta : \bar{\beta}+\bar{\beta} : \beta+\tau)+\bar{\beta} : (\beta : \gamma+\gamma : \beta+\tau)+\tau : \bar{\beta}+\tau : \gamma$$

■

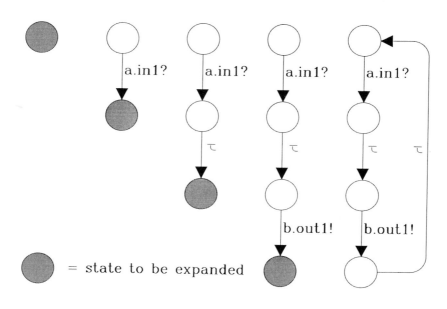

Figure 2.10: Illustrating the expansion law using state expansion.

2.7   Synchronization and value-passing

There are two versions of CCS; one which deals with pure synchronization and
one which deals with value passing as well. The first version can be considered
as a special case of the second version because a synchronization pulse can be
considered as a message without a value (like an envelope without a letter). Here,
we will focus on the version of CCS without value passing, i.e. we will focus on
pure synchronization. The reason for this is two-fold:

(a) it does not make sense to try to fully verify a system (including verification
    of the domains over which values may range) unless its *synchronization
    framework* is correct;

(b) it makes verifications easier when done by hand, and (in the case of many
    large expressions) requires less CPU time when done by a computer.

However, it should be mentioned that this apparent limitation is not really a fundamental one. It will serve our purpose and we will always be able to include values of messages by means of the following procedure. We will create a separate port for each message or message value which is considered to be relevant for the problem at hand. For instance, in the value-passing version of CCS, a term like *a.in?x* indicates that a message received via input port *a.in* will be assigned to variable *x*. In the pure synchronization case we have to interpret messages as attributes of port names. Hence, a term like *a.in?x* should be interpreted in the synchronization case as an action via port *a.inx*.

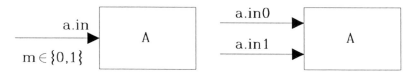

Figure 2.11: Translating message values into multiple ports.

Consider the following example. System A (*figure 2.11*) has an input port *a.in* via which a message *m* can be received; *m* can take the values 0 or 1. If we translate this to a situation where we only consider synchronization, then we have two choices: (i) we either omit the message and its possible range of values (we can do this in those situations in which we want to analyze the *effect* of a communication irrespective of the value transmitted); or (ii) we create a separate port for each of the relevant values of *m*. In *Figure 2.11*, we have created two ports *a.in0* and *a.in1* for the two values of *m*. In general, for each relevant message value we will simulate a separate port. Another way to treat value passing can be found in [Mil80]. We will use two naming schemes for ports:

(a) one with *names* and *co-names*, in which there is no explicit distinction between input ports and output ports. We will use Greek letters to indicate names and co-names. Communicating ports are represented by complementarity of names, such as in α and $\bar{\alpha}$.

(b) Alternatively, we will indicate explicitly which ports are input ports and which ports are output ports in cases where such distinction is required. Notice that in this case there is a distinction between the name of a port and the name of the corresponding action; we use a question mark following the name of an input port to denote an input action. For an output action we use an exclamation mark following the port's name.

Due to the way we will treat message values, we are able to use a syntax which can be used to express both pure synchronization as well as value passing:

$$
\begin{aligned}
<\text{action}> &\Rightarrow <\text{name}> [<\text{message}>] \\
&\Rightarrow <\text{co-name}> [<\text{message}>] \\
&\Rightarrow <\text{outport}>![<\text{message}>] \\
&\Rightarrow <\text{inport}>?[<\text{message}>] \\
&\Rightarrow \tau
\end{aligned}
$$

<name>, <co-name>, <outport>!, and <inport>? are elements from $\mathcal{L}$; <inport> is the name of an input port; <outport> is the name of an output port; <message> is received in case of an input port and sent in case of an output port. The message part of an action may be empty.

Chapter 3

Verification

3.1 The principle of verification in CCS

In CCS, verification is based on the definition of equivalence relations on agents. The main idea behind these equivalence relations is the notion of observation. Two agents are equivalent if an external observer, communicating with both of them, is unable to distinguish between the two agents. The type of equivalence relation that we will use is *bisimulation equivalence*. Roughly speaking, two agents are in bisimulation if each one can simulate the communication behaviour of the other. We will also introduce the notion of *observation equivalence*. Finally, we will summarize the laws derived by Milner. These CCS-laws maintain observation equivalence, and they can be used to transform one agent into another provided the two agents are observationally equivalent.

In *Section 1.1*, verification was defined as the activity in which the consistency between a specification and a potential implementation is checked. Using algebraic techniques like CCS, we can make this definition more formal in the following way:

(a) define an equivalence relation over behaviours;

(b) show that the behaviours of the specification S and the implementation I are in the same equivalence class.

In other words, if we cannot distinguish between S and I in this way, we may conclude that I correctly implements S. The behaviour of S is given in terms of a behaviour equation. Suppose that the designer comes up with a potential implementation, consisting of a decomposition of S in terms of interconnected agents, each again defined by means of a behaviour equation (*Figure 3.1*).

In order to verify the correctness of I with respect to S, we have to show that S and I have equivalent behaviours at the observable ports x, y and z of S. Bear in mind that the implementation I can be considered as a refinement of S. We will have to perform an abstraction step, removing these details from I;

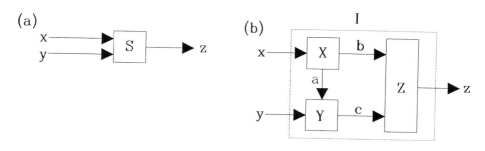

Figure 3.1: *A system S and a potential implementation I, consisting of the decomposition into subsystems X, Y, and Z.*

this enables us to express the behaviour of I at the level of S. In the example it means we have to remove certain details of the behaviour of I with respect to the communications via a, b and c. In fact, we only want to observe the *effect* of these communications at the external ports x, y and z. In summary we need to build an abstraction function A, such that:

$$\boxed{S = A(I)}$$

We shall refer to this as the **verification condition** (in *Sections 3.2* and *3.4* we will show what we mean by " = "). The abstraction function A consists of:

(i) concurrent composition, i.e. merging the behaviours of the subsystems using the expansion;

(ii) performing restrictions, i.e. hiding internal communications of I;

(iii) reducing the equations using the CCS laws.

Steps (i) and (ii) require the application of the expansion law. Hence, in our example we need to establish the truth of $S \approx (X \mid Y \mid Z)\backslash\{a, b, c\}$ using the CCS laws.

3.2   Bisimulation and observation equivalence

We want to analyze agents by communicating with them. The goal of this analysis is to find out whether two agents can be distinguished by an external observer. If an external observer cannot distinguish between agents P_1 and P_2, then we will say that P_1 and P_2 are equivalent. In this section we will give a definition of the type of equivalence that we will use. First, we will have a look at a few examples, where we consider agents that never perform τ-actions.

Case 1: Consider the agents P_1 and Q_1:

$$P_1 :: \quad P_1 = a : P_2 \qquad\qquad Q_1 :: \quad Q_1 = a : Q_2$$
$$ \quad P_2 = b : P_3 \qquad\qquad\qquad\quad\; Q_2 = b : Q_1$$
$$ \quad P_3 = a : P_2$$

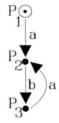

Both P_1 and Q_1 communicate via a and b consecutively, each starting with action a. The only difference between P_1 and Q_1 is in their number of states, but this cannot be observed externally. Therefore, P_1 and Q_1 are equivalent.

Case 2: Consider the agents R_1 and S_1:

$$R_1 :: \quad R_1 = a : R_2 + a : \text{NIL} \qquad S_1 :: \quad S_1 = a : S_2$$
$$ \quad R_2 = b : \text{NIL} \qquad\qquad\qquad\qquad\quad\; S_2 = b : \text{NIL}$$

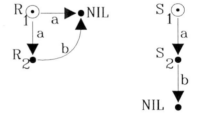

Initially, both R_1 and S_1 can only communicate via a. During the a-communication, R_1 makes an internal choice either to stop (i.e. to move to state NIL) or to permit a communication via b (i.e. to move to state R_2), whereas S_1 will always permit a b-communication after the initial a-communication. This difference can be observed by an external observer, and hence R_1 and S_1 are not equivalent.

Case 3: Consider the agents T_1 and U_1:

$$T_1 :: \quad T_1 = a : T_2 + a : T_3 \qquad U_1 :: \quad U_1 = a : U_2$$
$$ \quad T_2 = b : T_1 \qquad\qquad\qquad\qquad\quad\; U_2 = b : U_1 + c : U_1$$
$$ \quad T_3 = c : T_1$$

Initially, both T_1 and U_1 can only perform an a-communication. During the a-communication, T_1 makes an internal choice to move to state T_2, after which

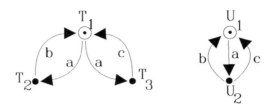

only a b-communication is possible, or to move to state T_3, after which only a c-communication is possible. U_1, however, will permit both b and c-communications, i.e. it leaves the choice to the environment. Hence, T_1 and U_1 are not equivalent.

The above examples did not contain τ actions, representing internal actions of agents. Next, let us see what happens if agents contain τ actions:

Case 4: Consider the agents X_1 and Y_1:

$$X_1 :: \quad X_1 = a : X_1 \qquad Y_1 :: \quad Y_1 = a : Y_1 + \tau : Y_2$$
$$Y_2 = a : Y_1$$

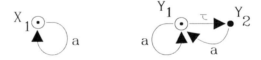

Both X_1 and Y_2 are able to perform an unbounded number of a-actions. The only difference between X_1 and Y_2 is that Y_2 can perform an internal action every now and then, while X_1 will never perform an internal action. Obviously, an external observer cannot distinguish between X_1 and Y_2.

Evaluating whether two agents are equivalent can be done on the basis of a definition of *bisimulation*. Before we can give its definition, we have to introduce a few other notions. We write $P \stackrel{\alpha}{\Longrightarrow} Q$ if $P(\stackrel{\tau}{\longrightarrow})^*(\stackrel{\alpha}{\longrightarrow})(\stackrel{\tau}{\longrightarrow})^*Q$, where $(\stackrel{\tau}{\longrightarrow})^*$ means zero or more τ's. By $\alpha \restriction$ we mean the sequence $(\stackrel{\tau}{\longrightarrow})^* \stackrel{\alpha}{\longrightarrow} (\stackrel{\tau}{\longrightarrow})^*$ with all τ's removed (including α, in the case where $\alpha = \tau$). We call Q an α-*derivative* of P if $P \stackrel{\alpha}{\longrightarrow} Q$. Likewise, we call Q an α-*descendant* of P if $P \stackrel{\alpha}{\Longrightarrow} Q$. For $(\stackrel{\tau}{\longrightarrow})^* \restriction$ we reserve the empty symbol ϵ (the empty action).

Definition 3.1 A relation $\mathcal{S} \subseteq \mathcal{E} \times \mathcal{E}$ over agents is called a **weak bisimulation** if $(P, Q) \in \mathcal{S}$ implies, for every $\alpha \in \mathcal{ACT}$:

(i)   whenever $P \stackrel{\alpha}{\longrightarrow} P'$ then, for some $Q', Q \stackrel{\alpha\restriction}{\Longrightarrow} Q'$ and $(P', Q') \in \mathcal{S}$

(ii)   whenever $Q \stackrel{\alpha}{\longrightarrow} Q'$ then, for some $P', P \stackrel{\alpha\restriction}{\Longrightarrow} P'$ and $(P', Q') \in \mathcal{S}$

A pair of agents P and Q is said to be **observation equivalent** (or *weakly bisimilar,*) written $P \approx Q$ if the following property holds (see also [Mil89], page 110-111):

> P and Q are observation equivalent if, for every action α, every α-derivative of P is observation equivalent to some α-descendant of Q, and similarly with P and Q interchanged.

In formula:

$P \approx Q$ if, for all $\alpha \in \mathcal{ACT}$,

(i)   whenever $P \xrightarrow{\alpha} P'$ then, for some $Q', Q \xRightarrow{\hat{\alpha}} Q'$ and $P' \approx Q'$

(ii) whenever $Q \xrightarrow{\alpha} Q'$ then, for some $P', P \xRightarrow{\hat{\alpha}} P'$ and $P' \approx Q'$

Milner showed that in order to prove $P \approx Q$ we only have to find a weak bisimulation which contains (P, Q) such that in each of the possible states P and Q can simulate one another.

If we look at our example agents X_1 and Y_1 again, we find that $\{(X_1, Y_1), (X_1, Y_2)\}$ is an appropriate weak bisimulation, and hence: $X_1 \approx Y_1$. This bisimulation can also be shown using state graphs (the dotted lines show states which are equivalent):

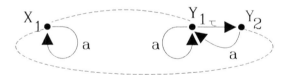

Consider the following behaviour trees:

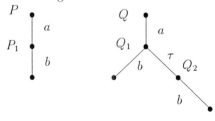

The question is: are P and Q observation equivalent? We translate this question into: is there a bisimulation that contains P an Q? So let us start with the pair (P, Q). We apply the definition in the two directions, i.e. we need to show that P and Q simulate one another. The next table shows the first step; *Direction*

refers to the two directions (i) and (ii) in the definition of bisimulation. The last column shows the pairs, for which bisimulation still needs to be shown.

Pair	Action	Direction	Derivate	Descendant	To be shown
(P,Q)	a	(i)	P_1	Q_1	(P_1,Q_1)
				or	or
				Q_2	(P_1,Q_2)
		(ii)	Q_1	P_1	(P_1,Q_1)

Both P and Q can only execute action a (and the sequence $a : \tau$ in case of Q). We conclude that in order for Q to simulate P, either P_1 and Q_1 or P_1 and Q_2 should be bisimular. Likewise, for P to simulate Q, we conclude that P_1 and Q_1 should be bisimular. Hence, showing P_1 and Q_1 to be bisimular is needed in both cases. We proceed with the pair (P_1, Q_1).

Pair	Action	Direction	Derivate	Descendant	To be shown
(P_1,Q_1)	b	(i)	NIL	NIL (twice)	(NIL,NIL)
		(ii)	NIL	NIL	(NIL,NIL)

Since equality implies equivalence, we conclude that for action b, P_1 and Q_1 simulate one another. However, the definition of bisimulation requires that we check simulation for each possible action. Since Q_1 can also perform a τ action, we need to proceed (notice that $\tau \models \epsilon$).

Pair	Action	Direction	Derivate	Descendant	To be shown
(P_1,Q_1)	ϵ	(i)	P_1	Q_2	(P_1,Q_2)
		(ii)	Q_2	P_1	(P_1,Q_2)

Hence, in order for (P_1, Q_1) to be in the bisimulation relation, we need to check (P_1, Q_2). The only action which can occur in these states is action b.

Pair	Action	Direction	Derivate	Descendant	To be shown
(P_1,Q_2)	b	(i)	NIL	NIL	(NIL,NIL)
		(ii)	NIL	NIL	(NIL,NIL)

Because equality implies equivalence, we conclude that $(NIL, NIL) \in S$. Hence, we conclude $P \approx Q$ since we have found a bisimulation which relates each state of agent P with at least one state of agent Q and vice versa. Or, stated otherwise, for each state of agent P there is a state in agent Q such that Q simulates P in that state and vice versa.

Exercise 3.1. Consider the handshake example of *Section 2.6* and prove:

$$C = (A \mid B)\backslash\{a,b\} \approx Z, \text{ where } Z = \text{ a.in1? : b.out1! : Z} \qquad \blacksquare$$

3.3   An algorithm for finding a weak bisimulation

In general, showing observation equivalence for a pair of agents consists of the following steps:

1. construction of the state graphs (or behaviour trees) of the two agents;

2. construction of a binary relation between the nodes of the two state graphs such that (a) the pair of initial nodes of the two state graphs is in the relation, 9b) each node is in at least one binary relation, and (c) the relation is a weak bisimulation.

The most inefficient algorithm to execute step 2 would be:

(i) Construct every possible binary relation between the node sets of the corresponding two state graphs, satisfying 2(a) and 2(b).

(ii) Check for each of these relations if it satisfies the conditions of a weak bisimulation. The two agents are observationally equivalent iff at least one of these relations is a weak bisimulation.

We will need a more efficient method. The following method is based on an approach developed by P.J. de Graaff [?].

Definition 3.2. For $X \in \mathcal{E}$ and $\alpha \in \mathcal{ACT}$ we define:

$d(X, \alpha) = \{X' \mid X \stackrel{\alpha}{\Longrightarrow} X'\}$, the set of all α-descendants of X.   ∎

Definition 3.3. For $i \geq 0$ we define the binary relation R_i as follows:

$$R_0 = \{(X, Y) \mid X \in \mathcal{E} \wedge Y \in \mathcal{E}\}$$
$$R_{i+1} = \{(X, Y) \in R_i \mid (\forall \alpha : \alpha \in \mathcal{ACT} : (\forall X' : X' \in d(X, \alpha) : (\exists Y' : Y' \in d(Y, \alpha) : (X', Y') \in R_i)) \wedge (\forall Y' : Y' \in d(Y, \alpha) : (\exists X' : X' \in d(X, \alpha) : (X', Y') \in R_i)))\}$$ ∎

One can find out whether two agents are observationally equivalent by computing $R_i, i \geq 0$ until one finds $R_i = R_{i+1}$, for some $i \geq 0$; then R_i is the required equivalence if each state of the agents X and Y appears at least in one pair of the relations R_i.

We can construct R_i in the form of a sequence of tables with the states of both agents alongside the rows and columns respectively; an entry 1 in the table for $R_i, i \geq 0$, means that the corresponding pair is in R_i. If a pair is not in R_i, then we write 0. From the definition of R_0 it follows that all entries in the table for

R_0 are equal to 1. The table for R_{i+1} can be obtained from the table for R_i and the table for $d(X, \alpha)$ in the following ways.

Let two agents X and Y have states X_j and Y_k respectively (j and k range over the number of states of X and Y respectively). Then $R_{i+1}(j, k) = 1$ if *(case (i))* for every $\alpha \in \mathcal{ACT}$ and X_l in $d(X_j, \alpha)$ there exists an Y_m in $d(Y_k, \alpha)$ such that $R_i(l, m) = 1$, and *(case (ii))* as case (i) with the roles of X and Y interchanged.

As an example we have another look at the agents P_1 and Q_1 *(Section 3.2 , Case 1)*; both can execute the following actions: ϵ (the empty action), a and b; if a system performs the empty action, then it either remains in the same state or the system moves to a new state via one or more τ actions. For the table of descendants we find:

$d(X, \alpha)$	$\alpha \setminus X$	P_1	P_2	P_3	Q_1	Q_2
	ϵ	$\{P_1\}$	$\{P_2\}$	$\{P_3\}$	$\{Q_1\}$	$\{Q_2\}$
	a	$\{P_2\}$	$\varnothing$	$\{P_2\}$	$\{Q_2\}$	$\varnothing$
	b	$\varnothing$	$\{P_3\}$	$\varnothing$	$\varnothing$	$\{Q_1\}$

From this table we derive:

R_0	P_1	P_2	P_3
Q_1	1	1	1
Q_2	1	1	1

R_1	P_1	P_2	P_3
Q_1	1	0	1
Q_2	0	1	0

R_2	P_1	P_2	P_3
Q_1	1	0	1
Q_2	0	1	0

So $\{(P_1, Q_1), (P_2, Q_2), (P_3, Q_1)\}$ is the required observation equivalence for the agents P_1 and Q_1.

For the agents R_1 and S_1 *(Section 3.2 , Case 2)* the descendant table is:

$d(X, \alpha)$	$\alpha \setminus X$	R_1	R_2	S_1	S_2
	ϵ	$\{R_1\}$	$\{R_2\}$	$\{S_1\}$	$\{S_2\}$
	a	$\{R_2, \text{Nil}\}$	$\varnothing$	$\{S_2\}$	$\varnothing$
	b	$\varnothing$	$\{\text{Nil}\}$	$\varnothing$	$\{\text{Nil}\}$

Exercise 3.2. Determine the bisimulation.   ■

3.4   Verification induction

Design usually is not an activity in which a final implementation is derived from the specification in a single step. In most cases the designer will work along a number of intermediate stages. This is the well-known hierarchy principle which allows the designer to tackle a large problem by replacing it with a set of smaller problems. This will be treated more fully in *Chapter 8*. It suffices here to say that the result of hierarchical design is a sequence of system models

$M_0, M_1, \ldots, M_n$

where M_0 is the initial specification; M_1 is a more detailed version of M_0, etc.
Finally we reach M_n, the final implementation. We can illustrate the step from
M_i to $M_j (0 \leq i < j \leq n)$ for the case where module S is decomposed as follows:

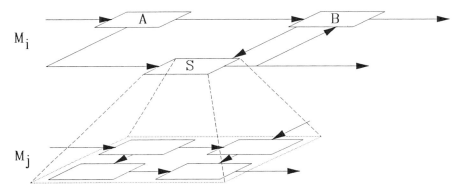

Figure 3.2: (a) a model M_i, and (b) the implementation of module S from M_i at
the level M_j (M_j also includes the implementations of modules A and B which
are not shown).

Figure 3.2 illustrates that each agent is itself decomposed into a set of communicating agents at the next refinement level. If we show equivalence between
these communicating agents and the 'higher-level' agents, we may conclude that
M_j is consistent with M_i if we show that M_j is equivalent with M_i. Hence, the
principle of **verification induction**:

(a) assume M_0 ;

(b) show M_{i+1} is equivalent with M_i for $0 \leq i <$n.

(c) then also M_n is equivalent with M_0. (This implies that the required equiv-
     alence is a transitive relation).

Hence, if we have verified correctness at each step, we may conclude that the
final implementation correctly implements the initial specification.

However, it appears that observation equivalence is too weak if we wish to apply
verification induction. The reason is that it is not sufficient for two agents to
be observation equivalent once we put those agents into another context. For
instance, suppose we have built a rack of equipment containing different printed
circuit boards (imagine S, A, and B from *Figure 3.2* to be these printed circuit
boards). Suppose we wish to replace one of these boards (say S) with another
board which has the same functionality but is constructed from new components
(i.e. the implementation I). In that case we wish the old rack and the new rack

to be observation equivalent. For the replaced printed circuit board we need a stronger equivalence, *observation congruence*, to account for equivalence within all possible contexts. If we put this in terms of CCS, we can distinguish between the following operational contexts for a system S:

S + T	(+-context)
S | T	(|-context)
S \\{a}	(\\-context)
α : S	(action context)
S[f]	(relabelling context)

We shall write $[\![S]\!]$ to denote S within its operational context, i.e. $[\![S]\!]$ is one of the above forms. Then the question is: If $S \approx I$, can we conclude $[\![S]\!] \approx [\![A(I)]\!]$?. In other words: if we replace an agent by another observation equivalent agent or set of agents, is the new system (containing the new agent(s)) observation equivalent with the previous system? In general, the answer is **no**! We need a stronger equivalence, which is also a congruence, i.e. which maintains equivalence within all operational contexts. It appears that observation equivalence is a congruence for all contexts, except the + context. More precisely, $B \approx D$ implies:

B | E	$\approx$	D | E
B\\{a}	$\approx$	D\\{a}
α :B	$\approx$	α : D
B[f]	$\approx$	D[f]
B + E	$\not\approx$	D + E (in general)

Definition 3.4.
B is observation congruent with D if for every expression context $[\![\ ]\!]$ we have

$$[\![B]\!] \approx [\![D]\!]$$ ∎

Milner shows that the following relation is the required observation congruence (See [Mil89] page 153):

Definition 3.5. (**Observation congruence**)
P and Q are observation congruent (we write $P = Q$), if for all α:

(i)   whenever $P \xrightarrow{\alpha} P'$ then, for some $Q', Q \xRightarrow{\alpha} Q'$ and $P' \approx Q'$
(ii) whenever $Q \xrightarrow{\alpha} Q'$ then, for some $P', P \xRightarrow{\alpha} P'$ and $P' \approx Q'$ ∎

Note that this definition only differs from that of $\approx$ in one respect: $\xRightarrow{\alpha}$ appears in place of $\xRightarrow{\hat{\alpha}}$. This means that each action of P or Q must be matched by *at least* one action of the other. Note also that this applies only to the first actions of P an Q; we require only $P' \approx Q'$, not $P' = Q'$. For example $a \approx \tau : a$, but $a \neq \tau : a$ since the left hand side cannot do the same action as the right hand side.

Under observation equivalence, a number of laws hold. These will be given in the next section. All laws under $\approx$ also hold under $=$, with one exception

$(B \approx \tau : B)$. In addition, a number of τ-*laws* hold under $+$; they play a very important role in verification.

Exercise 3.3. Show that $B \approx \tau : B$.   ■

Exercise 3.4. Show that from $B \approx \tau : B$ it does not follow that

$$B + E \approx \tau : B + E.$$ ■

3.5   CCS laws

Showing equivalence of two agents B and C can be done in three ways:

(i)   using the definition of observation equivalence;
(ii)   using the method presented in *Section 3.3*;
(iii) using a number of laws which can be used as rewrite rules.

In (iii), applying a rule transforms a behaviour equation into another equivalent equation. We shall list the laws and group them for each operator. In his first book [Mil80], Milner derived these laws by using stronger types of equivalence first and then weakening the equivalence. The purpose was to discover which laws hold under the stronger equivalence and which hold under weaker forms of equivalence. We shall not treat these equivalencies here, but only mention them for reference purposes. In descending order (i.e. from stronger to weaker) these are:

type of equivalence	symbol used by Milner
direct equivalence	$\equiv$
strong equivalence	$\sim$
observation congruence	$\approx^c$ or $=$
observation equivalence	$\approx$
With the property that: $\equiv$ implies $\sim$ implies $\approx^c$ implies $\approx$.	

We shall give a listing of all the laws which hold under observation congruence. For ease of notation, we will write "$=$" instead of "$\approx^c$". For each law, a reference is given to both books of Milner: For example, *Sum* $\equiv$ *1* is the first sum-law derived under direct equivalence [Mil80],(p.75); *Sum* refers to [Mil89], (proposition 3.1); *Com* refers to [Mil89], (proposition 3.8); $c3.7(3)$ refers to the $3rd$ rule under corollary 3.7 in chapter 3 of[Mil89]; $p3.9(1)$ is the first rule under proposition 3.9 in chapter 3, etc.

List of CCS laws	[Mil80] reference	[Mil89] reference
The SUM laws:		
1. $B_1+B_2= B_2+B_1$ (commutativity)	Sum $\equiv$ 1	Sum (1)
2. $B_1+(B_2+ B_3)=(B_1+ B_2)+B_3$ (associativity)	Sum $\equiv$ 2	Sum (2)
3. $B+Nil=B$ (NIL is unity of +)	Sum $\equiv$ 3	Sum (4)
4. $B+B=B$ (absorption)	Sum $\equiv$ 4	Sum (3)
The COMPOSITION laws:		
1. $B_1 \mid B_2=B_2 \mid B_1$ (commutativity)	Com $\sim$ 1	Com (1)
2. $B_1 \mid(B_2 \mid B_3)=(B_1 \mid B_2) \mid B_3$ (associativity)	Com $\sim$ 2	Com (2)
3. $B\mid Nil = B$ (NIL is unity of |)	Com $\sim$ 3	Com (3)
The RESTRICTION laws:		
1. $Nil\backslash$ a= Nil	Res $\equiv$ 1	
2. $(B_1+B_2)\backslash a=B_1\backslash a+B_2\backslash a$	Res $\equiv$ 2	c3.7 (3)
3. $(\alpha :B)\backslash a = Nil$ $(a=\alpha)$	Res $\equiv$ 3	c3.7 (1)
$= \alpha :(B \backslash a)$ $(a\neq \alpha)$		
4. $B\backslash a=B$ if $a\notin L(B)$ ($L(B)$ is the sort of B)	Res $\sim$ 1	p3.9 (1)
5. $B\backslash a\backslash b = B\backslash b\backslash a$	Res $\sim$ 2	p3.9 (2)
6. $(B_1 \mid B_2)\backslash a = (B_1\backslash a)\mid(B_2\backslash a)$	Res $\sim$ 3	p3.9 (4)
if $a\notin L(B1)\cap L(B_2)$		
The RELABELLING laws:		
1. $NIL[S] = NIL$	Rel $\equiv$ 1	
2. $(B_1 + B_2)[S] = B_1[S] + B_2[S]$	Rel $\equiv$ 2	c3.7 (4)
3. $(\alpha : B)[S] = S(\alpha) : B[S]$	Rel $\equiv$ 3	c3.7 (2)
4. $B[I] = B$ (I identity relabelling)	Rel $\sim$ 1	p3.10 (1)
5. $B[S] = B[S']$ provided $[S]$ and $[S']$	Rel $\sim$ 2	p3.10 (2)
have the same domain in $L(B)$.		
6. $B[S][S'] = B[S'\circ S]$	Rel $\sim$ 3	p3.10 (3)
("$\circ$" stands for function composition)		
7. $B[S]\backslash b = B\backslash a[S]$, where $S(a)=b$	Rel $\sim$ 4	p3.9 (3)
8. $(B_1 \mid B_2)[S]=B_1[S]\mid B_2[S]$	Rel $\sim$ 5	p3.10 (4)
The TAU-laws:		
1. $\alpha : \tau :B=\alpha :B$	$\tau - 1$	p3.2 (1)
2. $B+\tau :B=\tau :B$	$\tau - 2$	p3.2 (2)
3. $\alpha :(B+\tau :C)+\alpha :C=\alpha :(B+\tau :C)$	$\tau - 3$	p3.2 (2)
4. $B+\tau :(B+C)=\tau :(B+C)$	corr. 7.14	c3.3
And, under observation equivalence only:		
$B \approx \tau :B$	p7.1	p5.7

A complete axiomatization of $=$ requires the notion of *fairness*, to be introduced in *Section 4.1*.

3.6   Verifying large expressions

Problem:
given a set of agents, their behaviours and a connection table (e.g. as a block diagram), calculate the combined behaviour of the agents.

Solution:
Application of the expansion law will always work, but applying it in a slightly more intelligent way will simplify the effort. For instance, an expression like $(A \mid B \mid C \mid D \mid E)\backslash P$ (where P is the port set connecting A, B, C, D and E) can be expanded by selecting ABCDE as the initial state and applying the expansion law. However, suppose we see from the behaviours of B and C that they heavily interact. Then we could first calculate $(B \mid C)\backslash P$ and reduce the resulting equations. After that we would calculate the remaining compositions. Composing the behaviours of two strongly interacting agents, yields much simpler expressions after reduction as compared to composing the behaviours of weakly interacting agents. Hence, we can adopt a *"divide and conquer"* rule due to the fact that:

(a) parallel composition | is associative, and

(b) reducing equations using CCS-laws does not change their semantics.

Hence, the first thing we have to do is to determine the *order* in which we wish to calculate the expansion. For instance, we may observe that in the expression $(A \mid B \mid C \mid D \mid E)\backslash P$ the agents B and C, and the agents D and E strongly interact; this can be judged on the basis of the number of messages which these agents exchange (or the number of complementary labels). Then we would write $(A \mid (B \mid C) \mid (D \mid E))\backslash P$ to indicate that $(B \mid C)$ and $(D \mid E)$ are calculated first. We can put this **expansion order** in pictorial form as in *Figure 3.3(a)*. Alternatively, we may decide that $((A \mid (B \mid C)) \mid (D \mid E))\backslash P$ is the correct expansion order (*Figure 3.3(b)*).

From these simple examples we observe that some parallelism in the expansion is possible. In order to simplify the results of the partial expansions $(B \mid C)$ and $(D \mid E)$ using the CCS laws, we first have to perform restriction. To do so we need to distribute the port set P into the composition expression. If we have an expression $(A \mid (B \mid C) \mid (D \mid E))\backslash P$, then we can decompose P as follows:

$$P = \{P_{BC} \bigcup P_{DE} \bigcup P_{A(BC)} \bigcup P_{A(DE)} \bigcup P_{(BC)(DE)}\},$$

where P_{BC} is the port set connecting B and C; $P_{A(BC)}$ is the port set connecting A with the composite of B and C, etc. The decomposition has the property

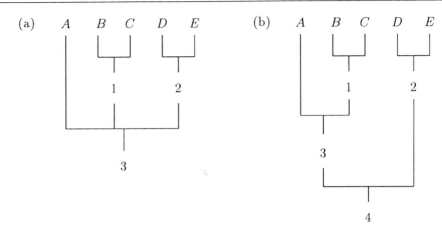

Figure 3.3: Illustration of the expansion order of (a) expression $(A \mid (B \mid C)_1 \mid (D \mid E)_2)_3 \backslash P$ (the sequence 1, 2, 3 indicates the sequence of expansions), and (b) expression $((A \mid (B \mid C)_1)_3 \mid (D \mid E)_2)_4 \backslash P$.

that $P_{BC} \cap P_{DE} = \varnothing$ (the same for any other pair). Then using rules (Res ~ 3)and$(\text{Res} \sim 1)$ we can write the above expression as:

$$(A \mid (B \mid C)\backslash P_{BC} \mid (D \mid E)\backslash P_{DE})\backslash \{P_{A(BC)} \bigcup P_{A(DE)} \bigcup P_{(BC)(DE)}\}$$

This shows how to distribute restriction over composition. Using rule (Rel~ 2) we can have a separate relabelling for each of the partial compositions. The procedure for expanding large compositions is as follows:

step 1: determine the expansion order, yielding several partial expansions (e.g. 1,2,3,4 in the above example).

Then for each partial expansion perform the following steps:

step 2: select a partial expansion and write the associated behaviours in normal form. If necessary, redefine behaviour identifiers for ease of writing (e.g. replace BCDE by X);

step 3: define a suitable relabelling, consistent with the connection table. (With "suitable" we mean that we can use simple names - like $a, b, c, \ldots$ - as the result of the relabelling. Since this relabelling is only defined within a specific context, we could use the same names in other contexts). Then apply the relabelling;

step 4: apply the expansion law;

step 5: reduce the resulting expressions using the CCS laws.

When reducing equations, one repeatedly applies the following types of operation:

(a) substitution,

(b) a CCS law.

Given a set of behaviour expressions, we give each expression a unique label. Substitution of an expression $e1$ into expression $e2$ is written as: $e1 \rightarrow e2$. Applying a CCS law $\mathbf{L}$ to an expression $e1$ is written as $\mathbf{L}(e1)$. For larger reductions, the following notation may be useful: $e1 \rightarrow e2, e3$ will mean: $e1 \rightarrow e2$ and $e1 \rightarrow e3$. Also, $e1, e2 \rightarrow e3$ will mean: $e1 \rightarrow e3$ and $e2 \rightarrow e3$.

Example:

$$
\begin{array}{ll}
A_1 = a?:A_2 + b?:A_3 & (1) \\
A_2 = c?:A_4 & (2) \\
A_3 = \tau :A_5 & (3) \\
A_4 = \tau :A_6 & (4) \\
A_5 = \tau :A_9 & (5) \\
A_6 = \tau :A_7 & (6) \\
A_7 = d!:A_8 + \tau :A_9 & (7) \\
A_8 = \tau :A_1 & (8) \\
A_9 = d!:A_1 & (9)
\end{array}
$$

As a general rule, reduction will proceed from the last expression towards the first expression. For the above example, we get:

$9 \rightarrow 7$	$A_7 = d! : A_8 + \tau : d! : A_1$	(7)
$9 \rightarrow 5$	$A_5 = \tau : d! : A_1$	(5)
$8 \rightarrow 7$	$A_7 = d! : \tau : A_1 + \tau : d! : A_1$	(7)
1τ (7)	$A_7 = d! : A_1 + \tau : d! : A_1$	(7)
2τ (7)	$A_7 = \tau : d! : A_1$	(7)
$7 \rightarrow 6$	$A_6 = \tau : \tau : d! : A_1$	(6)
$6 \rightarrow 4$	$A_4 = \tau : \tau : \tau : d! : A_1$	(4)
$4 \rightarrow 2$	$A_2 = c? : \tau : \tau : \tau : d! : A_1$	(2)
1τ (2) (3 times)	$A_2 = c? : d! : A_1$	(2)
$5 \rightarrow 3$	$A_3 = \tau : \tau : d! : A_1$	(3)
$3 \rightarrow 1$	$A_1 = a? : A_2 + b? : \tau : \tau : d! : A_1$	(1)
1τ (1) (2 times)	$A_1 = a? : A_2 + b? : d! : A_1$	(1)
$2 \rightarrow 1$	$A_1 = a? : c? : d! : A_1 + b? : d! : A_1$	(1)

The overall strategy for reduction is: from the last expression working towards the first expression apply the following operations repeatedly:

(a) find an expression which can be substituted into a second expression (or expressions); and then

(b) reduce the second expression using the τ-laws (and other laws).

3.7   An X-ray Diagnostic System (XDS)

Assume we have a system with which we can do X-ray diagnosis on patients. The system has an input and an output. Via the input we issue commands, via the output we observe (processed) X-ray images. Via *s.in* (*s.in* represents the keyboard of a terminal) the following commands can be given to the system:

XDS= s.in? exparams: s.out! image: XDS.
  + s.in? retrieve: s.out! image: XDS.

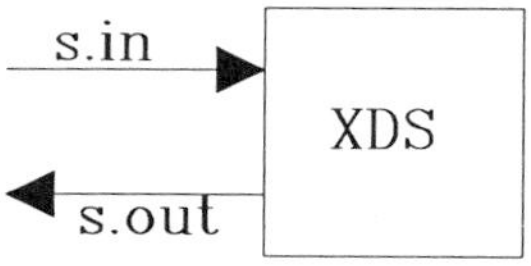

Figure 3.4: Specification of XDS.

exparams: This command starts an examination session. It indicates the voltage, pulse duration and energy to the X-ray system, together with other parameters. (We do not consider their values, since we concentrate on the synchronization framework).

retrieve: This command is used to retrieve a previously processed image. In our example we do not consider the parameters of this command (like the name of a patient).

Via *s.out* (which represents a TV-screen) we receive the required image. We assume that each newly obtained image is stored within the system and can be retrieved later. Hence, the system is capable of two alternative behaviours: processing and storing new images, or retrieving old ones. Suppose we make a global design and decompose the XDS system as indicated in *Figure 3.5* (observe that *s.in* is the same as the CONS input 1, and *s.out* is CONS output 4):

CONS: a console via which the operator interacts with the X-ray diagnostic system.

XRAY: the X-ray system comprising the X-ray generator.

PAT: the patient system. (patient, patient table X-ray detectors).

PROC: the image processor.

DISC: the storage system, where processed images are stored.

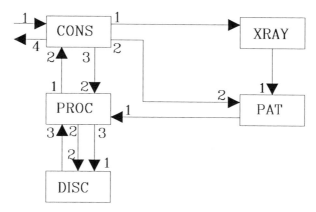

Figure 3.5: Decomposition of the X-ray system.

pos	position information for the patient table.
psd	process/store/display command for the image processor.
radiation	radiation produced by the X-ray tube.
xr.image	X-ray image from the X-ray detectors.
xr.params	X-ray parameters for the X-ray tube.
retrieve	retrieval command for images.
store	storage command for images.

Table 3.1: Internal messages of the XDS.

Between these subsystems the messages as indicated in *Table 3.1* are communicated.

Using the previously defined messages and the notational conventions, the equations for the subsystems are specified as follows:

CONS:: $C0 =$ *c.in*1? exparams: ; parameters for patient
                                                           ; examination are input from
                                                            ; the operator via console.
                     *c.out*1! xr.params: ; Next, parameters for X-ray
                                                            ; system are given
                                                            ; (voltage,duration, energy).
                     *c.out*2! pos: ; Then, position indication is
                                                            ; sent to patient system,
                     *c.out*3! psd:$C1$; and command for image
                                                            ; processor, indicating that
                                                            ; image has to be processed,
                                                            ; stored and displayed, is sent.
        $+$ *c.in*1? retrieve: ; Or the operator gives retrieve
                                                            ; instruction,

		$c.out3!$ retrieve:$C1$	; which is sent to processor.
	$C1 =$	$c.in2?$ image:	; Image is received from image
			; processor and is
		$c.out4!$ image:$C0$	; sent to display.
XRAY::	$X =$	$x.in?$ xr.params:	; X-ray parameters are received
			; from console and
		$x.out!$ radiation:X	; patient is X-rayed.
PAT::	$P =$	$p.in2?$ pos:	; Position parameters are
			; received from console and
		$p.in1?$ radiation:	; patient is X-rayed.
		$p.out!$ xr.image:P	; X-ray image is detected
			; by detector array.
PROC::	$I =$	$proc.in2?$ psd:	; image processor receives psd
			; command
		$proc.in1?$ xr.image:	; processor receives X-ray
		$process\ image$:	; image and processes it,
		$proc.out3!$ store:	; after which it instructs
		$proc.out2!$ image:	; the disc to store this
		$proc.out1!$ image:I	; image and sends the image to
			; console for display
	$+$	$proc.in2?$ retrieve:	; or processor receives
			; retrieve command, which it
		$proc.out3!$ retrieve:	; sends to the disc,
			; after which it receives the
		$proc.in3?$ image:	; image from the disc and
		$proc.out1!$ image:I	; sends it to the console
DISC::	$D =$	$d.in1?$ store:	; disc receives store
		$d.in2?$ image:	; instruction, and image,
		$store : D$	; which is stored
	$+$	$d.in1?$ retrieve:	; or disc receives retrieve
		$retrieve$:	; instruction, retrieves the
		$d.out!$ image:D	; image, and sends image to
			; console.

3.8 Verification of the XDS implementation

Following the reasoning of *Section 3.6* we shall first determine the expansion order. The expression we have to expand is:

(CONS|XRAY|PAT|PROC|DISC)$\backslash P$,

where $P = \{c.out1, c.out2, c.out3, c.in2, x.in, x.out, p.in1, p.in2, p.out, proc.in1,$

proc.in2, proc.in3, proc.out1, proc.out2, proc.out3, d.in1, d.in2, d.out}

An obvious choice is to compose PROC and DISC first. Another (arbitrary) choice is XRAY and PAT. Hence, we shall use the expansion order as depicted in *Figure 3.6.*

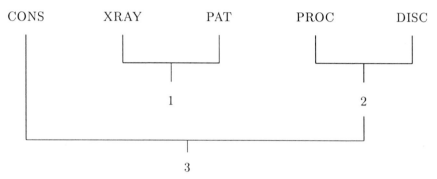

Figure 3.6: Expansion order for the X-ray diagnosticsystem XDS.

Putting the equations for XRAY and PAT in normal form and applying the following relabelling $x.out!$ radiation $\rightarrow a!$ and $p.in1?$ radiation $\rightarrow a?$ yields:

$$\text{XRAY:: } X = x.in? \text{ xr.params: } X_a \qquad \text{PAT:: } P = p.in2? \text{ pos: } P_a$$
$$X_a = a! : X \qquad\qquad\qquad P_a = a? : P_b$$
$$P_b = p.out! \text{ xr.image: } P$$

Hence, (XRAY | PAT)\ a yields:

$$XP = (x.in? \text{ xr.params} \mid p.in2? \text{ pos}): X_a P_a \qquad (\text{with } (c \mid d) = c : d + d : c)$$
$$X_a P_a = \tau : XP_b$$
$$XP_b = p.out! \text{ xr.image: } XP$$

Notice that XP_b does not have a $x.in?xr.params$ derivation since we assume that the operator will only issue such command at the start of the examination session. Using the $\tau - 1$-law we find:

$$XP = (x.in? \text{ xr.params} \mid p.in2? \text{ pos}): XP_b$$
$$XP_b = p.out! \text{ xr.image: } XP$$

Expansion 2 requires putting the equations for PROC and DISC in normal form and applying the following relabelling

proc.out2! image	$\rightarrow$	a!
proc.out3! store	$\rightarrow$	b!
proc.out3! retrieve	$\rightarrow$	c!
proc.in3? image	$\rightarrow$	d?
d.in1? store	$\rightarrow$	b?
d.in1? retrieve	$\rightarrow$	c?
d.in2? image	$\rightarrow$	a?
d.out! image	$\rightarrow$	d!

This yields:

$$PROC::\quad I \;=\; proc.in2?\ psd : I_a + proc.in2?\ retrieve: I_g$$

$$
\begin{aligned}
I_a &= proc.in1?\ xr.image : I_b & I_g &= c! : I_h \\
I_b &= process\ image : I_c & I_h &= d? : I_e \\
I_c &= b! : I_d & I_e &= proc.out1!image : I \\
I_d &= a! : I_e & &
\end{aligned}
$$

$$
\begin{aligned}
DISC::\quad D &= b? : D_a + c? : D_c \\
D_a &= a? : D_b & D_c &= retrieve : D_d \\
D_b &= store : D & D_d &= d! : D
\end{aligned}
$$

Exercise 3.5. Expand $(PROC|DISC)\backslash\{a, b, c, d\}$, and reduce: ID is initial state.
■

We find for $(PROC|DISC)\backslash\{a, b, c, d\}$:

$$
\begin{aligned}
ID \;=\; & proc.in2?psd : proc.in1?xr.image : process\ image : \\
& proc.out1!\ image : ID \\
& + proc.in2?retrieve : proc.out1!image : ID.
\end{aligned}
$$

Finally, for expansion 3, we put the equations for (CONS), (PROC|DISC) and (XRAY|PAT) in normal form, replace ID by X_0 and XP by Y_0, and apply the following relabelling:

c.out1! xr.params	$\rightarrow$	a!
c.out2! pos	$\rightarrow$	b!
c.out3! psd	$\rightarrow$	c!
c.out3! retrieve	$\rightarrow$	d!
c.in2? image	$\rightarrow$	e?
x.in1? xr.params	$\rightarrow$	a?
p.in2? pos	$\rightarrow$	b?
p.out! xr.image	$\rightarrow$	f!
proc.in2? psd	$\rightarrow$	c?
proc.in1? xr.image	$\rightarrow$	f?
proc.out1! image	$\rightarrow$	e!
proc.in2? retrieve	$\rightarrow$	d?

This yields:

CONS::
$$C_0 = \text{c.in1? exparams} : C_a + \text{c.in1? retrieve:} \ C_d$$
$$C_a = \text{a!} : C_b \qquad\qquad C_b = \text{b!} : C_c$$
$$C_c = \text{c!} : C_1 \qquad\qquad C_d = \text{d!} : C_1$$
$$C_1 = \text{e?} : C_1' \qquad\qquad C_1' = \text{c.out4! image} : C_0$$

XP::
$$X_0 = \text{a?} : X_1$$
$$\qquad + \text{b?} : X_2 \qquad\qquad X_1 = \text{b?:} \ X_3$$
$$X_2 = \text{a?} : X_3 \qquad\qquad X_3 = \text{f!} : X_0$$

ID::
$$Y_0 = \text{c?} : Y_1$$
$$\qquad + \text{d?} : Y_2 \qquad\qquad Y_1 = \text{f?} : Y_2$$
$$Y_2 = \text{e!} : Y_0$$

Exercise 3.6.
Verify that $(\text{CONS}|\text{XRAY}|\text{PAT}|\text{PROC}|\text{DISC})\backslash\{a,b,c,d,e,f\} \approx^c \text{XDS}$
assuming $s.in = c.in1$ and $s.out = c.out4$. ∎

3.9 OSI-layers

One would apply the following method in order to verify OSI-protocols. In the
OSI-model [Org84], layer 3 implements a reliable communication channel. This
is implemented on the basis of layer 2 and layer 1 protocols. Such protocols
should first be translated into CCS.

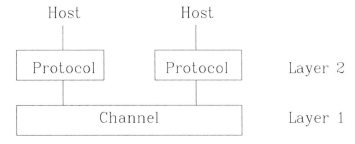

Figure 3.7: An OSI protocol structure.

A *Host* consists of a transmitter and a receiver sectionand communicates with
another *Host* via layers 1 and 2. Let *Protocol* denote the layer 2 protocol, and
let *Channel* denote the channel protocol of layer 1. Then verification requires
showing that:

$$(\text{Protocol}|\text{Channel}|\text{Protocol}|\text{Host})\backslash P \approx^m \text{Host}.$$

, where P denotes the internal actions.

Chapter 4

Fairness, queues and time

4.1 Fairness

In certain cases the CCS-laws presented in *Section 3.5* are not sufficient to build
a suitable abstraction function needed for verification *(Section 3.1)*. One such
case has to do with *fairness*; roughly speaking, the fairness property requires that
some unwanted behaviour will not last forever. We shall illustrate the necessity
for fairness using a suitable example: the alternating-bit protocol. Consider a
sender and a receiver, communicating via a channel:

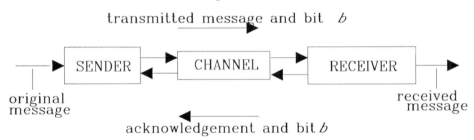

Figure 4.1: The alternating-bit protocol.

When offered a message, the sender adds a bit b to it and sends it via the
channel. The value of b is the complement of the value assigned to the previous
message. Hence, in a sequence of messages the value of b is alternating between
0 and 1; hence the name of the protocol. After sending the message and its
associated bit, the sender waits for the acknowledgement for the message; the
acknowledgement also carries a bit b which should have the same value as the
b value of the message. In case the b values of message and acknowledgement
differ, the sender retransmits the message.

Due to errors (e.g. noise) occurring on the channel, messages may be lost. To
detect this, a time-out is used. Whenever the time-out expires, the sender retransmits the message that was last sent.

49

When the correct acknowledgement is received, the sender returns to its initial
state and is now ready to accept a new message, to which it adds a bit with a
value complementary to the previously acknowledged message.

x	message accepted by sender from the data source;
y	message delivered by receiver to data sink;
d	message from sender to channel;
D	message from channel to receiver;
A	acknowledgement from channel to sender;
a	acknowledgement from receiver to channel.

Figure 4.2: Messages used in the example.

Figure 4.2 shows the messages that will be used to model the behaviours of
sender, channel and receiver. Notice that a connection *timeout* exists between
the channel and the sender. The reason is the following. Whenever the sender
sends a message to the channel, it also starts a time-out (*Figure 4.3a*).

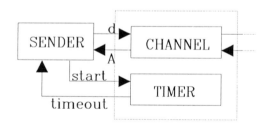

Figure 4.3: a) the actual situation; b) simplified situation.

Since actions *d* and *start* can be seen as a single action, we can simplify the
situation by abstracting the timer into the channel (*Figure 4.3b*); the timer is
started as soon as a message enters the channel. In *Section 4.3* we will consider
the treatment of time-outs in more detail.

Each message type comes in two versions (corresponding to the two values of the
b-bit), which we will model as two separate ports (remember how we use ports
and messages interchangeably as explained in *Section 2.7*). Hence we have ports
$x0$, $x1$, $y0$, $y1$, $d0$, $d1$, etc, where $x0$ reflects a message x to which a bit b with

value 0 has been attached. Likewise for the other messages. The specification of the whole system at x and y is:

$S = x0?\ :\ y0!\ :\ x1?\ :\ y1!\ :\ S$

For the sender we have:

$S_1 = x0?\ :\ S_5$
$S_2 = A0?\ :\ S_4 + A1?\ :\ S_5 + timeout?:S_5$
$S_3 = A1?\ :\ S_1 + A0?\ :\ S_6 + timeout?:S_6$
$S_4 = x1?\ :\ S_6$
$S_5 = d0!\ :\ S_2$
$S_6 = d1!\ :\ S_3$

And for the receiver:

$R_1 = D0?:R_5 + D1?:R_4$
$R_2 = a0!:R_3$
$R_3 = D1?:R_6 + D0?:R_2$
$R_4 = a1!:R_1$
$R_5 = y0!:R_2$
$R_6 = y1!:R_4$

We assume that the occurrence of a timeout is caused by the loss of an acknowledgement. This allows us to model the timeout within the channel. The timer will be started as soon as a message is offered to the channel, i.e. as soon as one of the actions $d0?$ or $d1?$ occurs. The event where the timer expires will be modeled as an action *timeout!*. This yields:

$C_1 = d0?:C_2 + d1?:C_3 + a0?:C_4 + a1?:C_5$
$C_2 = D0!:C_1 + timeout!:C_1$
$C_3 = D1!:C_1 + timeout!:C_1$
$C_4 = A0!:C_1 + timeout!:C_1$
$C_5 = A1!:C_1 + timeout!:C_1$

Exercise 4.1. Calculate $(S_1 \mid R_1 \mid C_1)\backslash\{a, A, d, D,\text{timeout}\}$ for each of the values of b; i.e. $a0, a1$, etc. ∎

The calculation yields (with renaming of the agent identifiers):

$(1)\quad S_1 = x0?\ :\ S_2$
$(3)\quad S_2 = \tau : y0!\ :\ S_3 + \tau : S_2$
$(6)\quad S_3 = \tau : x1?\ :\ S_5 + \tau : S_4$
$(10)\quad S_4 = \tau : S_3 + \tau : S_4$
$(11)\quad S_5 = \tau : y1!\ :\ S_6 + \tau : S_5$
$(14)\quad S_6 = \tau : S_1 + \tau : S_7$
$(16)\quad S_7 = \tau : S_6 + \tau : S_7$

At this point, we have exhausted the CCS-laws. Looking at equations (3), (10), (11) and (16), we observe that they are of the form:

$A = B + \tau : A$

This means that A can make τ-moves indefinitely. In our example this is caused by the occurrence of time-outs. This in turn was caused by the fact that messages can be lost within the channel. However, it seems fair to assume that some messages will get through undisturbed (otherwise the channel would be completely worthless!). This means that A above will not have infinite τ-moves, but will show B behaviour after a finite amount of τ-moves. Hence, the required inference rule for fairness is:

Fair
$$\frac{A = B + \tau : A}{A = \tau : B}$$

Bergstra and Klop have adopted this rule as KFAR (Koomen's Fair Abstraction Rule) in their Algebra of Communicating Processes [BK83, Bae86]. In their case, the rule is applied whenever the above situation occurs. However, here we will apply the rule only in those cases where the designer has additional information about the application such that he or she knows that the underlying implementation will guarantee that the rule is applicable. This is a design decision which is to be carried forward to subsequent design cycles. With reference to the rule $(\tau - 1)$, we can write *(Fair)* also as follows (τ^+ is a non-empty sequence of τ's):

Fair+
$$\frac{A = B + \tau^+ : A}{A = \tau : B}$$

Exercise 4.2. Show that application of *(Fair)* and other CCS laws, finally yields $S_1 = x0?:y0!:x1?:y1!:S_1$. ∎

We can generalize *(Fair)* to the case of n-mutual recursion. For $n = 2$ we have:

$$A = B + \tau : A + C$$
$$B = A + \tau : B + D$$

Let $[\![\,]\!]_{AB}$ denote the behaviour expression context in which A and B occur:

$[\![\,]\!]_{AB} ::$

$$A = B + \tau : A + C$$
$$B = A + \tau : B + D$$
... (rest of the expressions, involving A, B, and possibly other agents)

Let $[\![X/A, X/B]\!]_{AB}$ denote the behaviour context $[\![\,]\!]_{AB}$ in which each occurrence of A or B has been replaced by X. Then the second inference rule for fairness in the case of twofold mutual recursion is:

Fair-2
$$\frac{A = B + \tau : A + C,\ B = A + \tau : B + D,\ [\![\,]\!]_{AB}}{X = \tau : (C + D),\ [\![X/A, X/B]\!]_{AB}}$$

In other words: we replace the behaviour equations for A and B by the behaviour equation for X. At the same time, within the behaviour equations in which A and B occur, we replace the agent identifiers A and B by the agent identifier X. For example, consider the following behaviour context (we assume that A and B do not appear in the behaviour expressions for C and D):

before applying rule (Fair-2)	*after applying rule (Fair-2)*
A = B + τ : A + C	
B = A + τ : B + D	X = τ : (C + D)
C = ...	C = ...
D = ...	D = ...
E = τ : A	E = τ : X
F = a? : B	F = a? : X

Exercise 4.3. Write rule *(Fair-2)* in a similar way as *(Fair+)*. ■

The fairness rule generalizes in the obvious way:

Fair-n
$$\frac{A_i = A_1 + \ldots A_{i-1} + A_{i+1} + \ldots A_n + \tau : A_i + B_i, \ BE[\![]\!]_{A_1 \ldots A_n}}{X = \tau : \sum_i B_i, \ BE[\![X/A_i]\!]_{A_1 \ldots A_n}}$$

$(1 \leq i \leq n)$

4.2 FIFO communication

Up till now we have assumed that communication between two agents occurs simultaneously; sending by one agent coincides with a receiving action by the other agent. In other words, the channel connecting the two agents has no memory. What happens if this is no longer holds the case? That is, what happens if the sending agent puts a message into a channel and proceeds with its computation without waiting for the receiving agent? In that case taking the message from the channel by the receiving agent is a separate action. In the sequel we will define the derivation rules to deal with those communications between agents which involve channels showing FIFO behaviour.

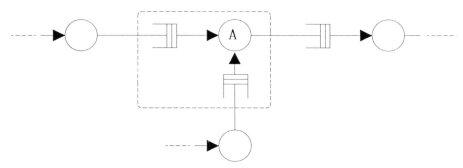

Figure 4.4: Queues connected to input ports.

In general we will assume that there exists an ordering relation between messages in a channel. Since we will be dealing with FIFOs, we assume this ordering to be a total ordering depending on the time of arrival at the channel. However, we are by no means restricted to FIFOs; one could assume other types of ordering relations (e.g. on the basis of a priority token associated with each message; a message with highest priority becomes available first at the other side of the channel). We will assume that the channel delay is negligible compared to delay that may occur within an agent itself. Hence, the channel delay is virtually zero.

In order to deal with the FIFO ordering of messages we will structure states in a certain way. Let S be an agent with n input ports. With each input port p we associate a message queue. We write $S(l)$ to denote the agent S with the message queue l at port p. $S(l1)(l2)\ldots(ln)$ shows the agent S with n queues. The message queue l consists of a list of messages $m_1 m_2 \ldots m_k$, where m_1 is the first message in the queue and m_k is the last message in the queue. We call l the **list of pending messages** in agent S for input port p. We define two functions h and t (from *head* and *tail* respectively):

$$h(l) = m_1 \quad t(l) = m_2 \ldots m_k$$
$$h() = \epsilon \quad t() = \epsilon \qquad (\epsilon \text{ is the empty list}).$$

By concatenating the head and tail of a queue, we obtain the original queue. We write this as $l = h(l), t(l)$. Since expansion is based on interleaving of actions (where we consider one action at a time), it is sufficient to consider the impact of a message on one particular message queue at a time.

With the above notation, expansion proceeds as follows. Suppose we have two agents A and B. A channel connects agent A with agent B. Whenever a message m is put on the channel by agent A, then this is modeled as a τ action for agent A, while at the same time message m is added to the list of pending messages associated with the corresponding input of agent B; we write $B(l)$ to include this queue for agent B.

In our inference rules for parallel composition with queues, expressions above the line are derivations, which are valid for the situation without queues. Below the line we will write derivations showing the effect of parallel composition with queues. Hence, if we know the behaviours of agents without queues, then these inference rules define parallel composition in those cases where queues should be taken into account. Dealing with the effect of queues during expansion is a more efficient method than representing queues as separate agents.

Like in *Chapter 2*, expressions above the line show the situation that should exist before the application of the rule, whereas the expressions below the line show the situation after the application of the inference rule. In the case of message transfer, we have to make explicit the direction of the transfer; λm means an input action λ involving message m (likewise for αm); the complementary action is an output action. In case $\alpha = \tau$, then $\alpha m = \tau m = \tau$. The rules at the left are valid in case queues are empty initially; the right hand rules show the general case:

$$\textbf{QCom}_1 \quad \frac{E \xrightarrow{\overline{\lambda m}} E', \ F \xrightarrow{\lambda m} F'}{E \mid F() \xrightarrow{\tau} E' \mid F(m)} \qquad \textbf{QCom}_1' \quad \frac{E \xrightarrow{\overline{\lambda m}} E', \ F \xrightarrow{\lambda m} F'}{E \mid F(l) \xrightarrow{\tau} E' \mid F(l,m)}$$

$$\textbf{QCom}_2 \quad \frac{E \xrightarrow{\lambda m} E', \ F \xrightarrow{\overline{\lambda m}} F'}{E() \mid F \xrightarrow{\tau} E(m) \mid F'} \qquad \textbf{QCom}_2' \quad \frac{E \xrightarrow{\lambda m} E', \ F \xrightarrow{\overline{\lambda m}} F'}{E(l) \mid F \xrightarrow{\tau} E(l,m) \mid F'}$$

$$\textbf{QCom}_3 \quad \frac{F \xrightarrow{\alpha m} F'}{F(m) \xrightarrow{\tau} F'()} \qquad \textbf{QCom}_3' \quad \frac{F \xrightarrow{\alpha m} F'}{F(m,l) \xrightarrow{\tau} F'(l)}$$

$$\textbf{QCom}_4 \quad \frac{F \xrightarrow{\alpha m} F'}{F() \xrightarrow{\alpha m} F(m)} \qquad \textbf{QCom}_4' \quad \frac{F \xrightarrow{\alpha m} F'}{F(l) \xrightarrow{\alpha m} F(l,m)}$$

Figure 4.5 illustrates the application of these rules $QCom_2$ has been omitted since it is analogous to $QCom_1$.

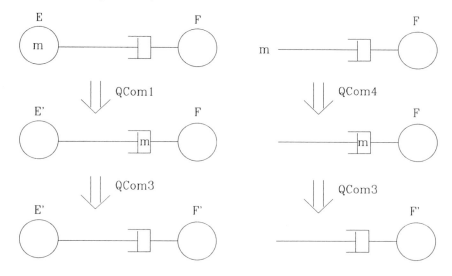

Figure 4.5: Illustration of inference rule.

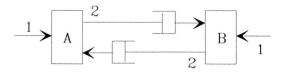

Figure 4.6: Agents A and B with two message queues.

Consider the handshake example of *Section 2.4*. Introducing messages and queues (*Figure 4.6*), we write the following behaviour equations for A and B:

$$
\begin{array}{llll}
\text{A} & = \text{a.in1?m : A1} & \text{B} & = \text{b.in?m : B1} \\
\text{A1} & = \text{a.out!m : A2} & \text{B1} & = \text{b.out1!m : B2} \\
\text{A2} & = \text{a.in2?ack : A} & \text{B2} & = \text{b.out2!ack : B}
\end{array}
$$

We associate one message queue with input b.in and a second message queue with input a.in2. Expanding $(A() \mid B())\backslash\{a.out1!m, b.out2!ack\}$, where a.out!m is the complement of $b.in?m$, and $b.out2!ack$ is the complement of $a.in2?ack$, yields:

$$
\begin{array}{lll}
\text{A()B()} & = \text{a.in1?m : A1()B()} & (\text{Com}_1) \\
\text{A1()B()} & = \tau : \text{A2()B(m)} & (\text{QCom}_1) \\
\text{A2()B(m)} & = \tau : \text{A2()B1()} & (\text{QCom}_3) \\
\text{A2()B1()} & = \text{b.out1!m : A2()B2()} & (\text{Com}_2) \\
\text{A2()B2()} & = \tau : \text{A2(ack)B()} & (\text{QCom}_2) \\
\text{A2(ack)B()} & = \tau : \text{A()B()} & (\text{QCom}_3)
\end{array}
$$

Or,

$$
\text{A()B()} \quad = \text{a.in1?} : \tau : \tau : \text{b.out1!} : \tau : \tau : \text{A()B()}
$$

Omitting m, the reduction yields: $A()B() = a.in1? : b.out1! : A()B()$, which is identical to the case without FIFO communication. This is an obvious result, since a handshake is supposed to be independent from internal delays. Notice that rule (QCom_4) could not be applied in view of the restriction. To simplify our notation still a step further, we may omit empty message queues. In the handshake example this would yield the following simplification:

$$
\begin{array}{ll}
\text{AB} & = \text{a.in1?m : A1B} \\
\text{A1B} & = \tau : \text{A2B(m)} \\
\text{A2B(m)} & = \tau : \text{A2B1} \\
\text{A2B1} & = \text{b.out1!m : A2B2} \\
\text{A2B2} & = \tau : \text{A2(ack)B} \\
\text{A2(ack)B} & = \tau : \text{AB}
\end{array}
$$

Without additional measures, introducing FIFO communication in general does not yield the same result after reduction. This can be observed in the following exercise.

Exercise 4.4. Consider the following agents A and B:

$$
\begin{array}{ll}
\text{A1} = \text{a : A2} & \text{B1} = b : \text{B2} \\
\text{A2} = \bar{b} : \text{A3} + \text{c : A3} & \text{B2} = \text{e : B3} \\
\text{A3} = \bar{b} : \text{A3} + \text{d : A1} & \text{B3} = \bar{d} : \text{B1}
\end{array}
$$

Calculate their combined behaviour with restriction over $\{b, d\}$; (i) in the normal case, and (ii) in the case where FIFO communication takes place between A and B. ∎

4.3   Deriving time-out conditions

4.3.1   A transmission protocol

Consider the following situation. A sender initiates a communication by sending a synchronization message SYN, followed by a length indicator LEN, a sequence number SEQ, an operation code OPC (indicating the way the data have to be used), the data itself (the number of bytes of which is determined by LEN), and finally a cyclic redundancy code CRC, to be used by the receiver to detect transmission errors. When all data have been correctly received, the receiver sends an acknowledgement ACK.

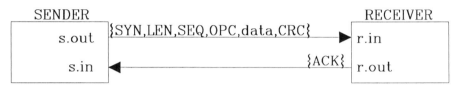

Figure 4.7: A sender, a receiver, and their messages.

We can model the above described behaviour in CCS as follows:

$$
\begin{array}{llll}
S_0 &=& \text{s.out!SYN} : S_1 & \qquad R_0 &=& \text{r.in?SYN} : R_1 \\
S_1 &=& \text{s.out!LEN} : S_2 & \qquad R_1 &=& \text{r.in?LEN} : R_2 \\
S_2 &=& \text{s.out!SEQ} : S_3 & \qquad R_2 &=& \text{r.in?SEQ} : R_3 \\
S_3 &=& \text{s.out!OPC} : S_4 & \qquad R_3 &=& \text{r.in?OPC} : R_4 \\
S_4 &=& \text{s.out!data} : S_4 & \qquad R_4 &=& \text{r.in?data} : R_4 \\
& & +\ \text{s\_enough} : S_5 & \qquad & & +\ \text{r\_enough} : R_5 \\
S_5 &=& \text{s.out!CRC} : S_6 & \qquad R_5 &=& \text{r.in?CRC} : R_6 \\
S_6 &=& \text{s.in?ACK} : S_0 & \qquad R_6 &=& \text{r.out!ACK} : R_0
\end{array}
$$

In S_4 and R_4 we assume that the sending/receiving of data is repeated a number of times defined by LEN. This implies that under error-free conditions both the sender and the receiver will decide that enough data have been sent or received at roughly the same time. This has been modeled by means of the actions s_enough and r_enough, of which the result is that sender and receiver proceed to states S_5 and R_5 respectively.

The above specification reflects the ideal case. In reality many errors can occur. In the sequel we will list the most probable errors. For each error it will be indicated how the specifications are affected and have to be changed to account for those errors. We will use the following syntax:

error\<number\>:	\<error description\>
decision:	\<effect on specification\>
equations:	\<effect on behaviour equations\>

error 1: The receiver detects a transmission error using CRC.

decision: In R_6 this has to be accounted for by introducing action $CRCerror$ after which the receiver will go to behaviour R_0. This means the sender will not receive ACK. To detect this the sender is supplied with a time-out. When this time-out occurs, we model this by the action s_tout.

equations: $S_6 = s.in?ACK : S_0 + s\_tout : S_0$ and
$R_6 = r.out!ACK : R_0 + CRCerror : R_0$.

(*The example is derived from an actual case and does not necessarily show the best design.*)

error 2: By introducing the time-out, the sender may decide that the receiver will not send ACK. However, the receiver may only be delayed.

decision: While in S_0, the sender must be able to receive ACK.

equations: $S_0 = s.out!SYN : S_1 + s.in?ACK : S_0$

error 3: An error may occur in the value of the length indicator LEN. We will distinguish between four error situations involving LEN.

error 3a: A transmission error causes the LEN value within the receiver to be too low. The receiver proceeds to R_5 and interprets the next data as CRC. However, the CRC check does not result in an error and, after sending ACK, the receiver proceeds to R_0. The sender, on the other hand receives an acknowledgement ACK while sending data or when ready to send CRC.

decision: While in R_0 the receiver must be able to receive data and/or CRC since the sending of these by the sender may coincide with the arrival of ACK at the sender. After receiving ACK, the sender proceeds to S_0.

equations: $S_4 = s.out!data : S_4 + s\_enough : S_5 + s.in?ACK : S_0$
$S_5 = s.out!CRC : S_6 + s.in?ACK : S_0$
$R_0 = r.in?SYN : R_1 + r.in?any : R_0$
(where *any* is any message other than SYN).

error 3b: The same as 3a but now the CRC check results in an error. The receiver proceeds to R_0. The sender finishes its normal transmission. Since no ACK will be received, time-out will occur.

decision: The equations already cover this.

error 3c: This time, a transmission error causes the *LEN* value to be too large at the receiver side. The sender finishes its normal sequence by sending *CRC*. The receiver interprets this as valid data. The sender will time-out since *ACK* is not received.

decision: Since the receiver is still in R_4 waiting for data (that will not arrive), an additional time-out is needed to bring the receiver to R_0.

equations: $R_4 = \text{r.in?data} : R_4 + r\_\text{enough} : R_5 + r\_\text{tout} : R_0$.

error 3d: Same as 3c but now *CRC* is not interpreted as valid data.

decision: The receiver proceeds to R_0 (the sender will time-out).

equations: $R_4 = \text{r.in?data} : (\tau : R_4 + \tau : R_0) + r\_\text{enough} : R_5 + r\_\text{tout} : R_0$. The expression $\tau : R_4 + \tau : R_0$ models the situation that, depending on the validity of the data, either R_4 is chosen or R_0; each τ expresses the two possible outcomes of the validity checking mechanism.

Exercise 4.5. With the above results, write the equations for sender and receiver in normal form. ∎

4.3.2 Calculating behaviour sequences.

In order to find out how the sender and receiver protocols interact, we will connect the sender and receiver together and calculate their combined behaviour. Hence, we are interested in the parallel composition of sender and receiver without taking into account the influence of the channel linking the two. Since we are interested in the occurrence of actions like *s_enough*, *r_enough*, *s_tout*, *r_tout* and *CRCerror*, we will do restriction over all other messages (for ease of writing, the restriction list below only shows these messages; the corresponding actions can be easily derived):

(sender|receiver)\{SYN,LEN,SEQ,OPC,data,CRC,ACK}

Exercise 4.6. Perform the above composition. You will find 23 behaviour equations in normal form. Analyse the resulting deadlock in $S_0 R_5$. ∎

We observe that the introduction of error behaviour and the use of time-outs has led to a possible deadlock situation! We can solve this particular problem by introducing a time-out in R_5:

$$R_5 \quad = \quad \text{r.in?CRC} : R_6 + \text{r\_tout} : R_0$$

We obtain the following behaviour equations for the sender and receiver:

$$
\begin{aligned}
S_0 &= \text{s.out!SYN} : S_1 & R_0 &= \text{r.in?SYN} : R_1 \\
&+ \text{s.in?ACK} : S_0 & &+ \text{r.in?any} : R_0
\end{aligned}
$$

$$
\begin{aligned}
S_1 &= \text{s.out!LEN} : S_2 & R_1 &= \text{r.in?LEN} : R_2 \\
S_2 &= \text{s.out!SEQ} : S_3 & R_2 &= \text{r.in?SEQ} : R_3 \\
S_3 &= \text{s.out!OPC} : S_4 & R_3 &= \text{r.in?OPC} : R_4 \\
S_4 &= \text{s.out!data} : S_4 & R_4 &= \text{r.in?data} : R'_4 \\
&\quad + \text{s\_enough} : S_5 & &\quad + \text{r\_enough} : R_5 \\
&\quad + \text{s.in?ACK} : S_0 & &\quad + \text{r\_tout} : R_0 \\
S_5 &= \text{s.out!CRC} : S_6 & R_5 &= \text{r.in?CRC} : R_6 \\
&\quad + \text{s.in?ACK} : S_0 & &\quad + \text{r\_tout} : R0 \\
S_6 &= \text{s.in?ACK} : S_0 & R_6 &= \text{r.out!ACK} : R_0 \\
&\quad + \text{s\_tout} : S_0 & &\quad + \text{CRCerror} : R_0 \\
& & R'_4 &= \tau : R_4 + \tau : R_0
\end{aligned}
$$

We calculate the combined behaviour of sender and receiver by evaluating the following expression (again, we only indicate the messages):

$$(S_0 \mid R_0) \backslash \{\text{SYN,LEN,SEQ,OPC,data,CRC,ACK}\}$$

We find:

$$
\begin{aligned}
S_0 R_0 &= \tau : S_1 R_1 & &(1) \\
S_1 R_1 &= \tau : S_2 R_2 & &(2) \\
S_2 R_2 &= \tau : S_3 R_3 & &(3) \\
S_3 R_3 &= \tau : S_4 R_4 & &(4) \\
S_4 R_4 &= \tau : S_4 R_4' & &(5) \\
&\quad + \text{s\_enough} : S_5 R_4 \\
&\quad + \text{r\_enough} : S_4 R_5 \\
&\quad + \text{r\_tout} : S_4 R_0 \\
S_4 R_4' &= \text{s\_enough} : S_5 R_4' & &(6) \\
&\quad + \tau : S_4 R_4 \\
&\quad + \tau : S_4 R_0 \\
S_5 R_4 &= \tau : S_6 R_4' & &(7) \\
&\quad + \text{r\_enough} : S_5 R_5 \\
&\quad + \text{r\_tout} : S_5 R_0 \\
S_4 R_5 &= \tau : S_4 R_6 & &(8) \\
&\quad + \text{s\_enough} : S_5 R_5 \\
&\quad + \text{r\_tout} : S_4 R_0 \\
S_4 R_0 &= \tau : S_4 R_0 & &(9) \\
&\quad + \text{s\_enough} : S_5 R_0 \\
S_5 R_4' &= \tau : S_5 R_4 & &(10) \\
&\quad + \tau : S_5 R_0 \\
S_6 R_4' &= \text{s\_tout} : S_0 R_4' & &(11) \\
&\quad + \tau : S_6 R_4 \\
&\quad + \tau : S_6 R_0 \\
S_5 R_5 &= \tau : S_6 R_6 & &(12)
\end{aligned}
$$

$$
\begin{aligned}
S_4 R_6 &= \text{s\_enough} : S_5 R_6 & &(14) \\
&\quad + \tau : S_0 R_0 \\
&\quad + \text{CRCerror} : S_4 R_0 \\
S_0 R_4' &= \tau : S_0 R_4 & &(15) \\
&\quad + \tau : S_0 R_0 \\
S_6 R_4 &= \text{r\_enough} : S_6 R_5 & &(16) \\
&\quad + \text{s\_tout} : S_0 R_4 \\
&\quad + \text{r\_tout} : S_6 R_0 \\
S_6 R_0 &= \text{s\_tout} : S_0 R_0 & &(17) \\
S_6 R_6 &= \tau : S_0 R_0 & &(18) \\
&\quad + \text{s\_tout} : S_0 R_6 \\
&\quad + \text{CRCerror} : S_6 R_0 \\
S_5 R_6 &= \tau : S_0 R_0 & &(19) \\
&\quad + \text{CRCerror} : S_5 R_0 \\
S_0 R_4 &= \text{r\_enough} : S_0 R_5 & &(20) \\
&\quad + \text{r\_tout} : S_0 R_0 \\
S_6 R_5 &= \text{s\_tout} : S_0 R_5 & &(21) \\
&\quad + \text{r\_tout} : S_6 R_0 \\
S_0 R_6 &= \tau : S_0 R_0 & &(22) \\
&\quad + \text{CRCerror} : S_0 R_0 \\
S_0 R_5 &= \text{r\_tout} : S_0 R_0 & &(23)
\end{aligned}
$$

$$S_5R_0 \quad = \quad \tau : S_6R_0 \qquad\qquad (13)$$

Note. Using rule (τ-1), we may write equation (1) as follows:

$$S_0R_0 = \tau : S_4\ R_4$$

4.3.3   The impact of timing conditions.

Using the above expansion, we are now able to analyse the impact of timing conditions. We will do this by looking at those equations from (1)-(20) that are affected by timing choices. By analyzing the possible action sequences in the above expansion; we will find that certain action sequences should not occur (these are indicated between brackets). From such constraints the required timing relations between actions can be derived.

(1) not affected.

$$
\begin{aligned}
(5)\ S_4R_4 \quad &= \quad \tau : S_4R_4\text{'} \\
&+ \text{ s\_enough} : S_5R_4 \qquad\quad \text{(sufficient data sent)} \\
&+ \text{ r\_enough} : S_4R_5 \qquad\quad \text{(sufficient data received)} \\
&(+\, \text{r\_tout} : S_4R_0) \qquad\quad \text{(receiver timer expires)}
\end{aligned}
$$

The time-out was included in R_4 to account for error 3c. However, this time-out should expire only after enough time has elapsed to allow data to be received or the action r_enough to occur. Let t(α) denote the time it takes for action α to occur. Then:

(a)   t(r_tout) > t(r_enough)
(b)   t(r_tout) > t(τ) = Δ

Where Δ is the time it takes to transmit data. In addition, we want:

(c)   t(r_enough) < Δ
        t(s_enough) < Δ

which tells us that the actions r_enough and s_enough should occur before new data are received. With these constraints, we can omit the last expression from equation (5) since at least one of the other action will have taken place by the time the timer runs out.

$$
\begin{aligned}
(6)\quad S_4R_4\text{'} = \quad & \text{s\_enough} : S_5R_4\text{'} \\
+\ & \tau : S_4R_4 \\
+\ & \tau : S_4R_0
\end{aligned}
$$

Since nothing can be said about the time when the decision whether the data is valid will be taken, the above equation remains intact.

$$
\begin{aligned}
(7)\quad S_5R_4 = \quad & \tau : S_6R_4\text{'} \qquad\quad (CRC \text{ interpreted as correct data}) \\
+\ & \text{r\_enough} : S_5R_5 \\
(+\ & \text{r\_tout} : S_5R_0)
\end{aligned}
$$

We can omit the last term since (a) and (b) still hold (receiver is in R_4).

$$
\begin{aligned}
\text{(8)} \quad S_4R_5 = \quad &\tau : S_4R_6 \qquad \text{(data read as } CRC) \\
&+ \quad \text{s\_enough: } S_5R_5 \\
&(+ \quad \text{r\_tout : } S_4R_0)
\end{aligned}
$$

The last action is prevented by (b) and (c) and can be omitted.

(9) not affected

(10) not affected; a similar reasoning as for (6) applies here.

(11) not affected; a similar reasoning as for (6) applies here.

$$
\begin{aligned}
\text{(12)} \quad S_5R_5 = \quad &\tau : S_6R_6 \qquad \text{(transferring the } CRC) \\
&(+ \quad \text{r\_tout : } S_5R_0)
\end{aligned}
$$

Preventing the time-out to occur here can be done by allowing sufficient time for the transmission of the CRC: this is already covered by (b) since Δ is the time it takes for the transfer of CRC.

(13) not affected.

(14) not affected.

(15) not affected.

$$
\begin{aligned}
\text{(16)} \quad S_6R_4 = \quad &\text{r\_enough : } S_6R_5 \\
&(+ \quad \text{s\_tout : } S_0R_4) \\
&+ \quad \text{r\_tout : } S_6R_0
\end{aligned}
$$

Since we wish the receiver to be reset before the sender we find:

(d) $t(\text{s\_tout}) > t(\text{r\_tout})$

On the basis of property (d) we can omit the second expression of S_6R_4.

(17)   not affected.

$$
\begin{aligned}
\text{(18)} \quad S_6R_6 \quad = \quad &\tau : S_0R_0 \qquad\qquad\qquad\qquad \text{(transfer of } ACK) \\
&+ \quad \text{CRCerror : } S_6R_0 \\
&(+ \text{s\_tout : } S_0R_6)
\end{aligned}
$$

The last action should not occur, since we want the required behaviour (involving ACK and $CRCerror$) to be possible. Usually $t(CRCerror) < \Delta$. Hence:

(e) $t(\text{s\_tout}) > \Delta$ (Δ is the time needed to transfer ACK).

and we may omit the last expression of S_6R_6.

(19)   not affected.

(20)   Due to our choice in (16), S_0R_4 cannot occur.

$$
\begin{aligned}
\text{(21)} \quad S_6R_5 \quad = \quad &(\text{s\_tout : } S_0R_5) \\
&+ \quad \text{r\_tout : } S_6R_0
\end{aligned}
$$

We wish the receiver to have the time-out before the transmitter does. This is guaranteed by condition (d). Hence, we may omit the first expression of $S_6 R_5$.

(22)   Due to our choice in (18), $S_0 R_6$ cannot occur.
(23)   not affected.

The results (a)–(e) have to be re-interpreted since the times we used were relative times. Since (a), (b) and (c) hold for the receiver while in state R_4, we do not have to change these. Hence, for the receiver, while in R_4, combining (a), (b) and (c):

(i) $t(\text{r\_enough}) < \Delta < t(\text{r\_tout})$

Equation (i) also covers (b), which should hold when the receiver is in R_5. Expression (d) defines an upper bound for $t(r\_tout)$ and is valid on two occasions; the receiver is either in state R_4 or state R_5, whereas the sender is in state S_6. We will analyse both situations. When the receiver is in R_4, the sender has progressed from S_4, via S_5 to S_6. The sender sends its last data word, which is received by the receiver after time Δ time units. The sender performs action s_enough, proceeds to S_5 and sends CRC, which, after Δ, is received by the receiver as a data word. Time-out r_tout is started. Sender proceeds to S_6 and starts s_tout.

We get:

(ii-a) $t(\text{s\_enough}) + t(\text{CRC}) + t(\text{s\_tout}) > 2\Delta + t(\text{r\_tout})$

where $t(s\_enough)$ and $t(CRC)$ are the times involved in the execution of s_enough and the sending of CRC respectively.

In case the receiver was in state R_5, then after receiving the sender's last data, the receiver waits for further data (remember, an error occurred in LEN). The receiver receives CRC and interprets this as valid data and (since in this case the value of LEN at the receiver is the value of LEN at the sender plus one) proceeds to R_5 and starts r_tout. Hence:

(ii-b) $t(\text{s\_enough}) + t(\text{CRC}) + t(\text{s\_tout}) > 2\Delta + t(\text{r\_enough}) + t(\text{r\_tout})$

which is a stronger condition for (s_tout) compared to (ii-a). For (e) we get the following: the sender starts s_tout upon entering S_6. The previously sent CRC is received by the receiver after time Δ. The receiver next prepares ACK and sends it. After time Δ the sender receives ACK. Hence, re-interpreting (e) yields:

(iii) $t(\text{s\_tout}) > 2\Delta + t(ACK)$

where $t(ACK)$ is the time needed by the receiver to send ACK.

Exercise 4.7. Carry out the above simplifications in equations (1)-(23) and reduce these equations using the CCS laws (*hint:* use fairness).   ∎

Exercise 4.8. Analyse the case where CRC is never interpreted as valid data (*hint:* investigate the consequences for expressions (14) and (19)).   ∎

4.4   Translating programs into CCS

4.4.1   The LAN protocol

We will show how to translate programs into CCS. As an example, we will treat
a simple protocol for a local area network. The original protocol from which this
example was derived was given in a procedural form. By translating the protocol
into CCS, we can use CCS for further analysis of the protocol.

Consider a number of stations $0 \ldots N$ connected to a unidirectional *ring*. Station
0 is the receiver; stations $1 \ldots N$ are senders. On the ring we have buckets which
can contain information. These buckets circulate on the ring. The following
algorithm guarantees mutual exclusion between the senders with respect to the
receiver. Each of the stations can perform the following basic operations for
accessing ring information.

read (a):	contents of a ring bucket are transferred to the variable a.
write (a):	a ring bucket receives the value of a.
swap (a):	the contents of a ring bucket and the value of a are interchanged.

Each of the buckets contain ternary variable (with values *empty, true* or *false*).

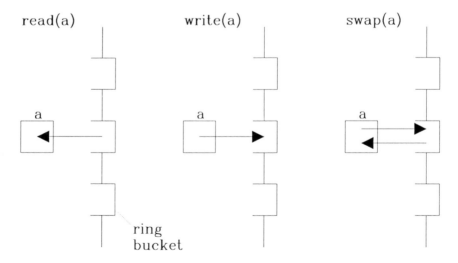

Figure 4.8: Illustration of the *read, write*, and *swap* operation. The buckets
constitute a ring-shaped LAN, and move one position at each tick of the clock.
Variable a is a local variable in one of the stations.

With these primitive operations, the following procedures are constructed:

insert(p) : **repeat** swap(p)   (initially: $p \neq$ empty)
           until $p \neq$ empty

wait(p)   : **repeat** read(a)   (initially: $p \neq$ empty)
           until $a = p$

With these procedures the following algorithms are constructed (they constitute the prologue of the critical section of sender and receiver):

SENDER   RECEIVER

p:= false; p:= true;
insert(p);   insert(p);
if not p **then** insert(p) **fi**;   **if** p **then** insert(p) **fi**;
while not p **if** p
do   **then**
 wait(true);   wait(false);
 insert(p)   insert(p)
od;   **fi**;

4.4.2   Some notational devices

(a) Procedural expansion without value-passing

If we wish to translate procedural expressions into CCS expressions, we need a mechanism to deal with procedure calls. The mechanism by which this will be done will be referred to as **procedural expansion**. It requires the introduction of two types of identifiers: (i) EXIT and (ii) "dotted" identifiers. For instance, consider the following behaviour equation:

   X = a : b : c : d

What would happen if the action c in fact itself was an agent identifier and c would be defined in terms of a behaviour equation? In terms of its behaviour tree, the leaves of the tree are either (i) agent identifiers referring to subtrees, or (ii) NIL. We will need a third type in order to model the return from a procedure: EXIT is a special agent identifier which refers to the context from which the current tree was reached. In our example, EXIT would refer to the colon following c in the behaviour equation for X.

Procedural expansion also requires to keep record of the context from which a certain behaviour expression was reached. We will treat the case of pure synchronization first and use the following notation to express context. Let S_1 and S_2 be agent identifiers, and let S_2 appear inside S_1. For example:

   $S_1 = a : b : S_2 : c \ldots + \ldots$

Then by $S_1 \cdot S_2$ we will mean that control has reached S_2 inside S_1; hence, we will remember context by means of these 'dotted' agent identifiers. Procedural expansion can be continued to arbitrary length. For example, suppose that in the above example, S_2 is defined as follows:

$S_2 = d : e : f : S_3 : g : $ EXIT and $S_3 = h : i : $ EXIT

Then $S_1 \cdot S_2 \cdot S_3$ indicates that control has reached S_3 within expression S_2 within S_1. In our example, procedural expansion could produce the following trace (states are inserted as references):

(S_1) a : b : $(S_1 \cdot S_2)$ d : e : f : $(S_1 \cdot S_2 \cdot S_3)$ h : i : g : c : ...

(b) Procedural expansion with value-passing

In addition to the use of EXIT and dotted identifiers, we will need a notational device to treat value passing. In particular, values have to be passed between subsequent expressions. Let S denote an agent identifier. Then

$S(a_1,\ldots,a_n ; b_1,\ldots,b_m)$

means that $a_1,\ldots,a_n$ are free variables within S which are bound to values received from the previous context. The variables $b_1,\ldots,b_m$ receive a value within S; these values are exported to the next expression (i.e. the one which follows S) where they bind to the associated free variables. For example,

$S(a_1, \ldots ,a_n ; b_1, \ldots , b_m) : S'(a'_1,\ldots ,a'_k ; b'_1 , \ldots ,b'_l)$

has the effect that the values of

$b_1, \ldots , b_m$ are bound to $a'_1 , \ldots ,a'_m$ for $m \leq k$ and
$b_1, \ldots , b_k$ are bound to $a'_1 , \ldots ,a'_k$ for $m \geq k$.

Consider the expression $S(a)$. It means that S imports a value from its context (the expression preceding S) and binds it to the free variable a inside S. Once a is bound to a value, $S(a)$ denotes a fixed agent identifier. In order to simplify the writing of expressions we will sometimes omit the export variables if no unambiguity arises; in that case we will assume that their values are bound to the import variable of the subsequent expression. For example, if $S(a)$ occurs in the following context:

$S(a;b) : S'(q)$

then we will write this simply as:

$S(a) : S'(q)$

Omitting the export variable from S simplifies the expression and implies that $S(a)$ yields a value to which q should be bound. In order to use this type of simplification, we have to identify an export variable inside the behaviour expression for S at the next lower level. For this purpose we will use the following notation inside the definition of S:

EXIT(q)

to indicate that the first free variable in the expression following S will be bound to q. For example, consider the expression

S' (;3) : S(a) : S"(q), where S is defined as:

$$S(a) \quad = \quad \textbf{if } a \geq 1000 \textbf{ then} \quad \text{out!a:EXIT}(1000)$$
$$+ \textbf{ if } a < 1000 \textbf{ then} \quad \text{EXIT}(a)$$

Then the above expressions yield $S"(3)$ once control reaches $S"$. This notation can be extended to any number of export values by identifying the position of the export values as follows: EXIT$(, q)$ and EXIT$(, , q)$ indicate that a value is bound to the second, respectively the third free variable in the subsequent expression. Whenever an export variable has been declared, like in $S(p; q)$, it receives the value *empty* initially.

4.4.3   Translation of the protocol into CCS

In this section we will put the LAN protocol introduced in *Section 4.4.1* in terms of CCS using the additional notation given in the previous section.

First, consider the expression for the *swap* operation. We will model it as an input from and output to the ring. The names of these are *in* and *out* respectively. Later, these names will be prefixed by the name of the subsystem to which they belong. For instance, in case of a sender s, the names become *s.in* and *s.out* respectively.

We will model *swap* as a sequence of an output and input action. In order to do so we will assume that the variable, the value of which will be swapped with the contents of a ring bucket, can be considered to consist of a read-variable p and a write-variable p. Both contain the same value (the value of the variable) until a new value is written into the write-variable. The value of the read-variable (which will be read and written into the bucket) is automatically updated to the value of the write-variable. Notice that the read-variable and write-variable are notational devices necessary to apply the CCS formalism.

Swap binds a free variable (the write-variable) and also exports a variable (the read-variable). The value of the latter is bound to the free variable in the next expression (in case there is no such free variable, the value of the export variable is maintained). For *swap* we then get:

$$
\begin{array}{lll}
\text{swap(p;q)} & = & \text{out!p:} \quad \text{the value of the read variable } p \text{ is} \\
& & \qquad\qquad \text{sent to a ring bucket.} \\
& & \text{in?q:} \quad \text{the value which was in this bucket} \\
& & \qquad\qquad \text{is bound to the write variable q.}
\end{array}
$$

EXIT(q)   after which we enter the next expression, while exporting q.

In expressions we will apply the simplification rule and write *swap(p)*. The *wait* procedure imports a value which is used for comparison against the contents of ring buckets. The same value is also exported, i.e. no computation is done which affects this value. For the *wait* procedure we get:

wait(p;p)   =   in?a : W(a,p;p)   read the content of a ring bucket.
W(a,p;p)   =   **if** a = p **then** EXIT(p)
           + **if** a $\neq$ p **then** wait(p;p)

Using the simplification rule we get:

wait(p)   =   in?a : W(a,p)
W(a,p)   =   **if** a = p **then** EXIT(p)
         + **if** a $\neq$ p **then** wait(p)

For the *insert* procedure we first write an equivalent expression:

insert(p):   swap(p);
                     if p = empty **then** insert(empty);

Insert has one import and one export variable. The previous expression can now be stated in terms of CCS as:

insert(p;q)   =   swap(p;q) : I(q)
I(q)   =   **if** q $\neq$ empty **then** EXIT(q)
I(q)   =   **if** q = empty **then** insert(empty).

Substituting swap and using the simplification rule, we get:

insert(p)   =   out!p : in?q : I(q)
I(q)   =   **if** q $\neq$ empty **then** EXIT(q)
           + **if** q = empty **then** insert(empty).

The *while-construct* "**while not** p **do** wait(true); insert(p) **od**;" yields:

WH(p;q)   =   **if not** p **then** wait(true) : insert(false;q) : WH(q)
          + **if** p **then** EXIT

We see that when *p* is true no variable is exported. When *p* is false, the value of variable *q* is exported to the next iteration of the while loop. Let *f* denote the assignment *p:=false*. The expression for the sender is:

S_0 =   f : insert(false):S_1(p)
S_1(p)   =   **if not** p **then** insert(false) : WH(p) : S_0
            + **if** p **then** WH(true) : S_0

S_0 is the initial agent identifier for the sender. Remember that insert exports a variable ; its value is bound to p in S_1. Let t denote the assignment $p:=true$. Then :

$$R_0 \quad = t : insert(true) : R_1(p)$$
$$R_1(p) = \textbf{if } p \textbf{ then} \quad insert(true) : R_2(p)$$
$$+ \textbf{ if not } p \textbf{ then} \quad R_2(false)$$
$$R_2(p) = \textbf{if } p \textbf{ then} \quad wait(false) : insert(true) : R_0$$
$$+ \textbf{ if not } p \textbf{ then} \quad R_0$$

4.4.4 Expansion of the sender and receiver expressions

First, we write all equations in normal form (*wait* is already in n.f.):

(a) $insert(p) \quad = \quad out!p : I$
  $I \quad\quad\quad = \quad in?q : I(q)$
  $I(q) \quad\quad = \quad \textbf{if } q \neq empty \textbf{ then} \quad EXIT(q)$
  $\quad\quad\quad\quad\quad + \textbf{ if } q = empty \textbf{ then} \quad insert(empty)$

(b) $WH(p) \quad = \quad \textbf{if not } p \textbf{ then} \quad wait(true) : w(p)$
  $\quad\quad\quad\quad\quad + \textbf{ if } p \textbf{ then} \quad EXIT$
  $w(p) \quad\quad = \quad insert(p) : WH(q)$

(c) $S_0 \quad\quad\quad = \quad f : S_{01}$
  $S_{01} \quad\quad\quad = \quad insert(false) : S_1(p)$
  $S_1(p) \quad\quad = \quad \textbf{if not } p \textbf{ then} \quad insert(false) : S_2(p)$
  $\quad\quad\quad\quad\quad + \textbf{ if } p \textbf{ then} \quad S_2(true)$
  $S_2(p) \quad\quad = \quad WH(p) : S_0$

(d) $R_0 \quad\quad\quad = \quad t : R_{01}$
  $R_{01} \quad\quad\quad = \quad insert(true) : R_1(p)$
  $R_1(p) \quad\quad = \quad \textbf{if } p \textbf{ then} \quad insert(true) : R_2(p)$
  $\quad\quad\quad\quad\quad + \textbf{ if not } p \textbf{ then} \quad R_2(false)$
  $R_2(p) \quad\quad = \quad \textbf{if } p \textbf{ then} \quad wait(false) : R_3(true)$
  $\quad\quad\quad\quad\quad + \textbf{ if not } p \textbf{ then} \quad R_0$
  $R_3(p) \quad\quad = \quad insert(p) : R_0$

Expanding the sender equations (c) using the expressions for *insert* (a) and *while* (b) yields:

$$S_0 \quad\quad\quad\quad\quad\quad\quad\quad\quad\quad = \quad f : S_{01}$$
$$S_{01} \quad\quad\quad\quad\quad\quad\quad\quad\quad = \quad s.out!false : S_{01}.i$$
$$S_{01}.i \quad\quad\quad\quad\quad\quad\quad\quad\quad = \quad s.in?true : S_{01}.I(true)$$
$$+ \quad s.in?false : S_{01}.I(false)$$
$$+ \quad s.in?empty : S_{01}.I(empty)$$

$S_{01}.I(true)$ $\quad\quad\quad = \quad \tau : S_1(true) \quad$ (binding)

$S_{01}.I(false)$ $\quad\quad\quad = \quad \tau : S_1(false)$

$S_{01}.I(empty)$ $\quad\quad\quad = \quad \tau : S_{01}.insert(empty)$

(no ambiguity since we are one level deep inside *insert*).

$S_1(true)$ $\quad\quad\quad = \quad \tau : S_2(true)$

$S_1(false)$ $\quad\quad\quad = \quad \tau : S_1.insert(false)$

$S_{01}.insert(empty)$ $\quad\quad\quad = \quad$ s.out!empty : $S_{01}.i$

$S_2(true)$ $\quad\quad\quad = \quad \tau : S_0$

$S_1.insert(false)$ $\quad\quad\quad = \quad$ s.out!false : $S_1.i$

($insert(false)$ omitted since i unambiguously indicates the location).

$S_1.i$ $\quad\quad\quad = \quad$ s.in?true : $S_1.I(true)$

$\quad\quad\quad + \quad$ s.in?false : $S_1.I(false)$

$\quad\quad\quad + \quad$ s.in?empty : $S_1.I(empty)$

$S_1.I(true)$ $\quad\quad\quad = \quad \tau : S_2(true)$

$S_1.I(false)$ $\quad\quad\quad = \quad \tau : S_2(false)$

$S_1.I(empty)$ $\quad\quad\quad = \quad \tau : S_1.I(empty).insert(empty)$

$S_2(false)$ $\quad\quad\quad = \quad \tau : S_2(false).wait(true)$

$S_1.I(empty).insert(empty)$ $\quad\quad\quad = \quad$ s.out!empty : $S_1.i$

$S_2(false).wait(true)$ $\quad\quad\quad = \quad$ s.in?a:$S_2(false).wait(true).W(true,a)$

(p = *false* persistent)

$S_2(false).wait(true).W(true,a)$ $\quad\quad\quad = \quad$ **if** a = true **then** $\quad S_2.W(false)$

$\quad\quad\quad + \quad$ **if** a $\neq$ true **then** $\quad S_2(false).wait(true)$

$S_2.W(false)$ $\quad\quad\quad = \quad$ s.out!false : $S_2.i$

$S_2.i$ $\quad\quad\quad = \quad$ s.in?true : $S_2.I(true)$

$\quad\quad\quad + \quad$ s.in?false : $S_2.I(false)$

$\quad\quad\quad + \quad$ s.in?empty : $S_2.I(empty)$

$S_2.I(true)$ $\quad\quad\quad = \quad \tau : S_2.WH(true)$

$S_2.I(false)$ $\quad\quad\quad = \quad \tau : S_2.WH(false)$

$S_2.I(empty)$ $\quad\quad\quad = \quad \tau : S_2.insert(empty)$

$S_2.WH(true)$ $\quad\quad\quad = \quad S_2(true)$

$S_2.WH(false)$ $\quad\quad\quad = \quad S_2(false)$

$S_2.insert(empty)$ $\quad\quad\quad = \quad$ s.out!empty : $S_2.i$

After substitution, reduction and renaming of agent identifiers we get for the *Sender*:

$S_0 \quad = \quad$ f : s.out!false : S_1

$S_1 \quad = \quad$ s.in?true : S_0

$\quad\quad + \quad$ s.in?false : s.out!false : S_2

$\quad\quad + \quad$ s.in?empty : s.out!empty : S_1

$S_2 \quad = \quad$ s.in?true : S_0

$\quad\quad + \quad$ s.in?false : s.in?a : $S_3(a)$

$\quad\quad + \quad$ s.in?empty : s.out!empty : S_2

$$S_3(a) \quad = \quad \textbf{if } a = \text{true } \textbf{then} \quad \text{s.out!false} : S_4$$
$$+ \ \textbf{if } a \neq \text{true } \textbf{then} \quad \text{s.in?a} : S_3$$
$$S_4 \quad = \quad \text{s.in?true} : S_0$$
$$+ \ \text{s.in?false} : \text{s.in?a} : S_3$$
$$+ \ \text{s.in?empty} : \text{s.out!empty} : S_4$$

Evaluating the expression for *Receiver* yields:

R_0	$=$	$t : R_{01}$
R_{01}	$=$	$\text{r.out!true} : R_{01}.i$
$R_{01}.i$	$=$	$\text{r.in?true} : R_{01}.I(\text{true})$
	$+$	$\text{r.in?false} : R_{01}.I(\text{false})$
	$+$	$\text{r.in?empty} : R_{01}.I(\text{empty})$
$R_{01}.I(\text{true})$	$=$	$\tau : R_1(\text{true})$
$R_{01}.I(\text{false})$	$=$	$\tau : R_1(\text{false})$
$R_{01}.I(\text{empty})$	$=$	$\tau : R_{01}.\text{insert}(\text{empty})$
$R_1(\text{true})$	$=$	$\tau : R_1.\text{insert}(\text{true})$
$R_1(\text{false})$	$=$	$\tau : R_2(\text{false})$
$R_{01}.\text{insert}(\text{empty})$	$=$	$\text{r.out!empty} : R_{01}.i$
$R_1.\text{insert}(\text{true})$	$=$	$\text{r.out!true} : R_1.i$
$R_2(\text{false})$	$=$	$\tau : R_0$
$R_1.i$	$=$	$\text{r.in?true} : R_1.I(\text{true})$
	$+$	$\text{r.in?false} : R_1.I(\text{false})$
	$+$	$\text{r.in?empty} : R_1.I(\text{empty})$
$R_1.I(\text{true})$	$=$	$\tau : R_2(\text{true})$
$R_1.I(\text{false})$	$=$	$\tau : R_2(\text{false})$
$R_1.I(\text{empty})$	$=$	$\tau : R_1.\text{insert}(\text{empty})$
$R_2(\text{true})$	$=$	$\tau : R_2.\text{wait}(\text{false})$
$R_1.\text{insert}(\text{empty})$	$=$	$\text{r.out!empty} : R_1.i$
$R_2.\text{wait}(\text{false})$	$=$	$\text{r.in?a} : R_2.W(\text{false},a)$
$R_2.W(\text{false},a)$	$=$	$\textbf{if } a = \text{false } \textbf{then} \quad R_3(\text{true})$
	$+$	$\textbf{if } a \neq \text{false } \textbf{then} \quad R_2.\text{wait}(\text{false})$
$R_3(\text{true})$	$=$	$\text{r.out!true} : R_3.i$
$R_3.i$	$=$	$\text{r.in?true} : R_3.I(\text{true})$
	$+$	$\text{r.in?false} : R_3.I(\text{false})$
	$+$	$\text{r.in?empty} : R_3.I(\text{empty})$
$R_3.I(\text{true})$	$=$	$\tau : R_0$
$R_3.I(\text{false})$	$=$	$\tau : R_0$
$R_3.I(\text{empty})$	$=$	$\tau : R_3.\text{insert}(\text{empty})$
$R_3.\text{insert}(\text{empty})$	$=$	$\text{r.out!empty} : R_3.i$

After substitution and renaming of agent identifiers we get for the *Receiver*

R_0	$=$	$t : \text{r.out!true} : R_1$
R_1	$=$	$\text{r.in?true} : \text{r.out!true} : R_2$
	$+$	$\text{r.in?false} : R_0$

$$
\begin{array}{rl}
& + \ \text{r.in?empty} : \text{r.out!empty} : R_1 \\
R_2 \quad = & \ \text{r.in?true} : \text{r.in?a} : R_3(a) \\
& + \ \text{r.in?false} : R_0 \\
& + \ \text{r.in?empty} : \text{r.out!empty} : R_2 \\
R_3(a) \quad = & \ \textbf{if } a = \text{false } \textbf{then} \ \ \text{r.out!true} : R_4 \\
& + \ \textbf{if } a \neq \text{false } \textbf{then} \ \ \text{r.in?a} : R_3(a) \\
R_4 \quad = & \ \text{r.in?true} : R_0 \\
& + \ \text{r.in?false} : R_0 \\
& + \ \text{r.in?empty} : \text{r.out!empty} : R_4
\end{array}
$$

The above expressions can now be used to analyze the protocol further, e.g. on deadlock, or to calculate the combined behaviour of sender and receiver. The reader is also referred to [Tan88], which contains several protocol algorithms that could be analysed using the method presented here.

Chapter 5

Synthesis

5.1 Synthesis strategies

In *Section 1.1* the basic design cycle was introduced. The cycle contains two major activities: (i) *verification* of potential implementations against a specification (this was the subject of *Chapter 3*), and (ii) *synthesis* or the creation of potential implementations. The synthesis activity requires most of the creative potential of the designer; as a consequence, synthesis is prone to errors. It may not yield the optimal result (or even a correct one) when carried out with too little experience or when the designer is dealing with very complicated problems. Due to its nature, synthesis is also very difficult to automate. Usually, automation of synthesis activities takes place "from the bottom upwards"; let us consider the situation for semi-custom IC design.

Lay-out synthesis was the first step to be taken; much of effort has gone into finding suitable algorithms for the placement and routing of the specified set of standard cells onto the chip area. *Symbolic lay-out editors* [MC80], e.g. based on stick diagram, or using procedural lay-out rules were the next step. *Logic synthesis* [Bea84, Tre87] aimed at the synthesis of logic functions in terms of simpler functions (AND, OR, etc.). Logic synthesis is used to automatically generate netlists for (parts of) VLSI circuits. A next step is *behavioural synthesis* in which the behaviour of a certain type of system, such as a digital signal processor [dMRC86] is mapped onto a basic architecture for such systems, followed by performance optimization. When designing a VLSI circuit, one encounters roughly the following design stages (we assume there exists a library of standard cells or macro cells):

(*i*) specifying the behaviour of the VLSI circuit;

(*ii*) (the global design stage) refining and decomposing that behaviour yielding a structure of high level, usually abstract, modules;

(*iii*) (the detailed design stage) decomposing further and mapping onto physically realizable or existing building blocks (standard cells).

In case the required building blocks do not exist, a next stage involving the design of cells, is required:

(*iv*) implementing the required building blocks (this requires detailed knowledge of the silicon process to be used).

Each stage is followed by a verification of the result by means of inspection, high level simulation, detailed simulation, breadboarding or formal verification. Generally speaking, synthesis (e.g. of a VLSI circuit or a software program) involves *refinement of behaviour* and the *decomposition of behaviour into structure* (in *Chapter 8* we will treat this more extensively). If we stay within the formal framework of CCS, synthesis would mean refining and decomposing behaviour equations into more detailed equations, such that the new equations would be observationally equivalent (or congruent) with the original equations. One can distinguish the following types of synthesis:

(*a*) high-level synthesis involving stages (*i*) and (*ii*);

(*b*) low-level synthesis involving stages (*iii*) and (iv);

(*c*) behavioural synthesis, mainly characterized by a large complexity of behaviour rather than complexity of the corresponding structure;

(*d*) structural synthesis, where behaviour of each element in the implementation may be relatively simple, but the complexity arises from the intricacy of the connections between the elements.

Usually, behavioural synthesis is synonymous with high-level synthesis, whereas low-level synthesis mainly corresponds with structural synthesis. Here, we will focus on high level synthesis of (*i*); instead of an VLSI implementation we could also decide to use a software implementation. We are interested in synthesis procedures for a class of systems which we will refer to as **interface systems** (such as communication protocols; see also [Mil87]). Such systems are characterized as follows: given a set of systems (or subsystems, or modules), find a new system which interfaces with the other systems in a predefined way. After the synthesis of an interface system in this way, one could use types (*a*)-(d) for further implementations. We will treat three types of interface synthesis:

(a) *synthesis by mirroring;*

(b) *synthesis by interface derivation;*

(c) *synthesis by completing the specification.*

5.2   Synthesis by mirroring

5.2.1   The principle of mirroring

Given an agent A. We want to derive an agent S which communicates with A without the danger of deadlock. More specifically, we want S to be such that $C(A, S) = \tau : C(A, S)$. Where $C(A, S)$ is a short-hand notation for $(A \mid S) \backslash L$ and L is the set of complementary actions of A and S, i.e. L represents the communication between A and S.

Figure 5.1: Given an agent A, find an agent S such that (i) S communicates with A, and (ii) $C(A, S) = \tau : C(A, S)$.

If A does not contain τ's, then the above requirement is certainly the case if $S = M(A)$, where $M(A)$ is the **mirror image** of A and is defined as follows:

$$M(A) = A[\overline{\alpha}/\alpha], \qquad\qquad \text{for all } \alpha \in L(A),$$

i.e. each action in A is replaced by its complementary action (e.g. an action a is replaced by $\overline{a}$, an action $\overline{b}$ is replaced by b, etc.). In case A contains τ's in a "+" context (i.e. equations of the form $\tau : X + Y$), then we obtain the **tau-reduced mirror image** $MT(A)$:

$$MT(A) = A\,[\overline{\alpha}/\alpha, /\tau]$$

in which actions are replaced by complementary actions, and τ's are removed.

Exercise 5.1. Determine the tau-reduced mirror image of

$$\tau : (\alpha + \tau : \beta) + \gamma$$

∎

There is one additional problem that we need to address; that of non-determinism. Consider the following behaviour of an agent A:

    A0 = a.in?m : A1 + τ : A2
    A1 = a.out1! : A0
    A2 = a.in?m : A3
    A3 = a.out2!m : A0

If we calculate $MT(A) = A[\overline{a}/\alpha, /\tau]$, we would get:

M0 = a.in!m : M1 + a.in!m : M3
M1 = a.out1? : M0
M3 = a.out2?m : M0

If we now calculate $C(A, MT(A))$, then this would yield:

A0M0 = τ : A1M1 + τ : A1M3 + τ : A2M0
A1M1 = τ : A0M0
A1M3 = NIL
A2M0 = τ : A3M1 + τ : A3M3
A3M1 = NIL
A3M3 = τ : A0M0

And we would have deadlocks in $A1M3$ and $A3M1$. This is certainly not what we would expect. The problem lies in the fact that, while in $M0$, system M can select any of the two actions $a.in!m$, leading to either $M1$ and $M3$. The solution lies in giving each occurrence of this action a unique identification. For example, we could give the first occurrence of this action the additional identifier 1, the second occurrence 2, etc. If we give the occurrences of the actions of $MT(A)$ unique identifiers, then we obtain the **tau-reduced mirror image with unique action occurrences** of agent A, abbreviated as MTU(A). Applying this to the above example, would yield:

A0 = a.in?m-1 : A1 + τ : A2   M0 = a.in!m-1 : M1 + a.in!m-2 : M3
A1 = a.out1! : A0   M1 = a.out1? : M0
A2 = a.in?m-2 : A3   M3 = a.out2?m : M0
A3 = a.out2!m : A0

After calculating and reducing $C(A,MTU(A))$, we find $A0M0 = \tau : A0M0$. We will illustrate the steps of *synthesis by mirroring* using a larger example. Suppose we wish to design a test system(*Figure 5.2a*). It is intended to test another system, while this latter system is in operation. Within the system to be tested, certain devices have a test interface. The test system itself comprises a microcomputer and a number of test circuits $A, B, \ldots, X$. The microcomputer and circuits are connected via a bus. Each circuit is connected to a particular test device via the test interface of the device. Each circuit (*Figures 5.2b, 5.3*) is composed of a device interface, a status register, a receiver (receiving messages from the micro computer), and a transmitter (sending messages to the micro-computer). Given the design of circuit A, what is the required program to be executed by the microcomputer for that particular circuit (and likewise for the other circuits $B, \ldots, X$)?

STEP 1: calculate the behaviour of the system with which S should interface

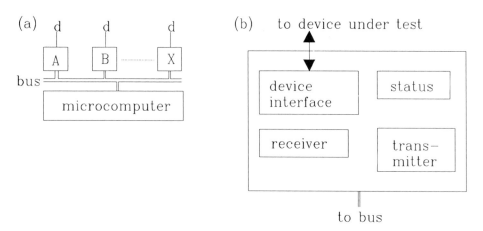

Figure 5.2: (a) architecture of the test system comprising a microcomputer and circuits $A, \ldots, X$; (b) constituents of one of the circuits $A, \ldots, X$.

We assume that we have defined the behaviour of each circuit in terms of the behaviours of its components (receiver, transmitter, status register and device interface). When we have defined these, we calculate the circuit behaviour by means of parallel composition. In order to keep the example simple enough, while still reflecting a real design, the behaviours of the device interface and the status register will be accounted for in the behaviours of the receiver and transmitter. Hence, we will only use these latter behaviours for our calculations. The behaviour of the receiver is given by the following set of behaviour equations:

R0 = r.in?sb :	;receiver waits for a start bit
r.out1!dis :	;then sends a disable signal
r.in?b : R1	;and receives a byte of data
R1 = τ : r.out1!en : R0	;depending on the contents of the
	;byte (modeled by the τ)
+ τ : r.out1!att : R2	;a choice is made
+ τ : R0	;to send enable or attention signal
	;or receiver decides to re-initialize
R2 = τ : r.out1!RD : R0	;data RD are sent
+ τ : R3	;or receiver decides to move to R3
R3 = τ : R5	;internal action causes a move to R5
+ τ : r.out1!RR : R0	;or receiver decides to send data RR
+ τ : r.out1!nack : R0	;or sends a negative acknowledgement
R5 = r.out1!ack : R0	;finally, an acknowledgement is sent

The behaviour of the transmitter is given by the following equations:

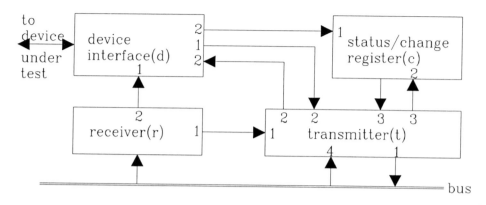

Figure 5.3: Wiring diagram: r.out1 = $\overline{t.in1}$, etc.

TE = t.in1?dis : TD	;the transmitter is disabled
+ τ : t.out1!m1 : T1	;or it sends m1 after internal move
+ t.in4?busy : TD	;or a busy signal is received
TD = t.in1?en : TE	;the transmitter is enabled
+ t.in1?dis : TD	;any further disabling is discarded
+ t.in1?att : t.out1!m2 : T0	;after receiving an attention signal,
	;the transmitter sends the
	;m2 message
T0 = t.in1?RD : T2	;data RD are received
+ t.in1?RR : T3	;or RR
+ t.in1?ack : T4	;an acknowledgement may be received
+ t.in1?nack : T4	;or a negative acknowledgement
T1 = t.in4?m1 : TD	;receiving m1 triggers a reset action
T2 = t.out1!0D : T4	;in T2, T3 and T4 several messages
T3 = t.out1!0R : T4	;are exchanged, reflecting the
T4 = t.out1!0ts : TD	;status of the device under test

Since we wish to derive the behaviour of the microcomputer towards a circuit, we need to know the communication between a circuit and the microcomputer. Therefore we calculate the behaviour of a circuit towards the bus. This requires the following parallel composition:

CIRCUIT = C(transmitter,receiver)

STEP 2: calculate the mirror image

After performing the required composition, followed by reducing the resulting equations using the CCS laws, we calculate *MTU(CIRCUIT)* and obtain the following result:

M0 = r.in!sb-1 : M1+r.in!sb-2 : M3+t.out1?m1-a : M5+tin4!busy-1 : M2

M1 = t.out1?m1-b : M6+tin4!busy-2 : M4+r.in!b : M7
M2 = r.in!sb-3 : M4
M3 = t.out1?m1-b : M6
M4 = r.in!b : M7
M5 = r.in!sb-4 : M6+tin4!m1-a : M2
M6 = tin4!m1-b : M4
M7 = r.in!sb-1 : M1+r.in!sb-2 : M3+t.out1?m1-a : M5
      + tin4!busy-1 : M2+t.out1?m2-b : M10+t.out1?m2-e : M13
      + t.out1?m2-d : M12+t.out1?m2-c : M9+t.out1?m2-a : M8
      + r.in!sb-3 : M4
M8 = r.in!sb-5 : M11+t.out1?0D-1 : M12+r.in!sb-7 : M15
      + t.out1?0R-1 : M12+r.in!sb-6 : M14+t.out1?0ts-1 : M2
M9 = r.in!sb-7 : M15+t.out1?0R-1 : M12
      + r.in!sb-6 : M14+t.out1?0ts-1 : M2
M10= r.in!sb-5 : M11+t.out1?0D-1 : M12
M11= t.out1?0D-2 : M14
M12= r.in!sb-6 : M14+t.out1?0ts-1 : M2
M13= in!sb-7 : M15+t.out1?0R-1 : M12
M14= t.out1?0ts-2 : M4
M15= t.out1?0R-2 : M14

STEP 3: refine the obtained program further

Assuming the micro computer has a multi-tasking operating system and a memory-mapped I/O, then the next steps could be:

STEP 3 − i:

Copy the above program for each of the circuits connected to the bus or make the code for the above program re-entrant;

STEP 3-ii:

Replace each input or output by the appropriate read or write operation to memory. For example $r.in!sb$ should be interpreted as a write operation to memory at location $r.in!sb$; after this operation, the corresponding memory location would then contain the value sb. $WRITE(r.in!sb,sb)$ would be an example of an operation performing this. Likewise, $t.out1?m1$ is the read operation from memory location $t.out1?m1$ to obtain the value for $m1$: $READ(t.out1?m1 , m1)$ would be the required read action of the micro computer.

5.2.2   Mirror observation equivalence

Suppose we have a system A, of which the behaviour can be written as:

$$A = \sum_i (\tau^* : \alpha_i : A_i) + \sum_j (\beta_j : A_j) + \tau : A + \tau : NIL. \qquad (\alpha_i \neq \tau, \beta_j \neq \tau)$$

The A_i and A_j can be written in a similar manner. The above is the most general form of behaviour equation that we can have (i.e. a behaviour equation has this general form, or a subset of it). We also define $X = \sum_i (\alpha_i : A_i) + \sum_j \beta_j : A_j$. Using fairness, and rule *Sum 4* it is easy to show that $X = MTU(A)$.

Let $CMTU(A) = C(A, MTU(A))$. It can be shown that

$$CMTU(A) = \tau : (\tau : CMTU(A) + \tau : NIL).$$

Exercise 5.2. Show this.   ∎

Suppose A would not have any sub behaviour of the form $\tau : NIL$. In that case we would have: $CMTU(A) = \tau : CMTU(A)$ as we have seen in the case of agent A in *Section 5.2.1*.

Exercise 5.3. Show
(a) $C(A, B) = \tau : C(A, B)$, where

$$A = \tau : \alpha : A1 + \beta : A1 \qquad A1 = \bar{\gamma} : A + \bar{\delta} : A$$
$$B = \bar{\alpha} : B1 + \bar{\beta} : B1 \qquad B1 = \tau : \gamma : B + \delta : B;$$

(b) $CMTU(\text{CIRCUIT}) = \tau : CMTU(\text{CIRCUIT})$; for CIRCUIT, see *Section 5.2.1*.
                                                                              ∎

We can generalize the above result by making use of the following symmetry. Whenever a system can perform a τ action when it is in a certain state, then its mirror image should not be able to do so. However, the roles of A and X at each step of the expansion $CMTU(A)$ can be reversed; the only thing that matters is that at each step one agent (A or X) should be the MTU of the other. Hence, we would find that $C(A, X) = \tau : C(A, X)$ also in the case where the roles of A and X would be interchanged at each expansion step.

Agents A and X that have the above type of behaviour will be called **mirror observation equivalent** . We can give a definition for this new equivalence, which we will denote as $\approx^m$ as follows:

A relation $\mathcal{S} \subseteq \mathcal{E}x\mathcal{E}$ over agents is called a **weak mirror bisimulation** if $(P, Q) \in \mathcal{S}$ implies, for every $\alpha \in \mathcal{ACT}$:
Definition 5.1

   (i) whenever $P \xrightarrow{\alpha} P'$ then, for some $Q', Q \xRightarrow{\hat{\alpha}} Q'$ and $(P', Q') \in \mathcal{S}$

   (ii) whenever $Q \xrightarrow{\alpha} Q'$ then, for some $P', P \xRightarrow{\hat{\alpha}} P'$ and $(P', Q') \in \mathcal{S}$

   (iii) whenever $P \xrightarrow{\tau} P'$ then $\neg(\exists Q', Q \xRightarrow{\tau} Q')$ and $(P', Q) \in \mathcal{S}$

   (iv) whenever $Q \xrightarrow{\tau} Q'$ then $\neg(\exists P', P \xRightarrow{\tau} P')$ and $(P, Q') \in \mathcal{S}$

Like in the case of observation equivalence, we need the largest weak mirror bisimulation to get mirror observation equivalence, relating every state of one agent with a state of the other and vice versa.

5.3 Synthesis by interface derivation

5.3.1 Causal relations

Very often during system design we are confronted with the following situation: we are given a number of modules and we need to design an interface linking the modules together in such a way that each module can express its behaviour in the right way.

Figure 5.4: Given agents A and B, find agent S such that (i) S communicates with A and B, and (ii) $C(A, B, S) = \tau : C(A, B, S)$. The general case is: given agents $A, B, \ldots$ find S such that $C(A, B, \ldots, S) = \tau : C(A, B, \ldots, S)$.

For example, signaling protocols are used to link telephone exchanges together in order to establish connections between subscribers. Since several such protocols are in use, two different protocols may have to be linked via a protocol converting interface. Another example is a microprocessor for which we have to design an interface for driving a printer. In general, given agents $S_1 \ldots S_n$, we want to derive an interface agent S that communicates with $S_1 \ldots S_n$ without the danger of deadlock. More specifically, we want S to be such that

$$C(\text{S}, S_1, \ldots, S_n) = \tau : C(\text{S}, S_1, \ldots, S_n). \text{ (for } 1 \leq i \leq n)$$

The latter is the case if

$$C(\text{S} \mid \text{S}_1 \mid \ldots \mid \text{S}_{i-1} \mid \text{S}_{i+1} \mid \ldots \mid \text{S}_n) \approx^{\text{m}} \text{S}_i,$$

Since we will need to consider all S_i, we will call the set $\{\text{MTU}(S_1), \ldots, \text{MTU}(S_n)\}$ the **partial specification** of S. We will use the symbol Sp to denote this partial specification.

In certain cases (which we will come back to), we will have to remember which messages have been received by the interface system. For this purpose we will associate a queue with the interface system. This means that we will add a list of messages to the agent identifier of the interface system S. However, we will assume the inputs of the interface system not to have queues associated with them (although one may add these later if required, using the method of *Section 4.2*). Also, we will assume that the internal delay of the interface system is negligible compared to the communication with the other systems (in *Chapter 9* we will consider an example where the synthesis is done on the basis of the assumption that the interface system has internal delay).

In addition to the above assumptions, we will need to define certain relationships that exist between messages at the different interfaces between S and the systems S_i. A **Causal relation** between an action α and an action β, written as $\alpha = \; - > \beta$ means that the occurrence of α is a necessary condition for β to occur. If action α has taken place, then action β can take place as soon as the agent of β has reached the state where it can execute β. We define two **causality predicates** L-CAUSAL( ) and R-CAUSAL( ) as follows:

L-CAUSAL(α) is true whenever the action α occurs at the left-hand side of a causal relation. R-CAUSAL(β) is true whenever the action β occurs at the right-hand side of a causal relation.

5.3.2 Synthesis rules

Next, we will derive a set of synthesis rules which describe the possible action derivations of S, i.e. the behaviour of S. Let us first introduce some notation to ease the writing:

* let S denote the current state of system S;

* let S' denote the next state of system S;

* let $S(l)$ denote the state of S; in addition, a list l is associated with the state of S; it contains the messages which have been received by S, and for which the causality predicate is true.

* IN(S) is the set of input actions that are possible when S is in its current state; depending on whether these input actions appear in causal relations, the associated messages may have to be stored in l.

* OUT(S) is the set of output actions that are possible when S is in its current state; depending on whether these output actions appear in causal relations, the actual occurrence of these actions depends on the availability of the corresponding messages in l.

Let $in(m_a)?$ denote an input action involving message m_a at some interface between S and $S_i (0 < i \leq n)$. Likewise, let $out(m_b)!$ denote an output action involving m_b at some other interface between S and $S_j (0 < j \leq n)$.

A **synthesis rule** will be written as a inference rule. The hypothesis part summarizes the conditions that should hold prior to the application of the rule. The

conclusion part then tells us what will be the situation after the application of
the rule. Let $h(l)$ and $t(l)$ denote the head and tail functions respectively, defined
on the queue l.

$$\textbf{Synth-1} \qquad \frac{\alpha \in \text{IN}(S), \text{L-CAUSAL}(\alpha), S_p \xrightarrow{\alpha} S'_p}{S(l) \xrightarrow{\alpha} S'(l, m)} \qquad\qquad (\alpha = \text{in}(m)?)$$

This rule tells us that if an input action, involving a message m, appears in a
causal relation within the context of the synthesis of interface system S, and if
the partial specification of S has a derivation from Sp to Sp' due to this action,
then the interface system S also has this derivation and the message m is put
into the queue.

$$\textbf{Synth-2} \qquad \frac{\alpha? \in \text{IN}(S), \neg\text{L-CAUSAL}(\alpha), S_p \xrightarrow{\alpha} S'_p}{S(l) \xrightarrow{\alpha} S'(l)} \qquad\qquad (\alpha = \text{in}(m)?)$$

This rule is very similar to the previous one, except that in this case the message
m is not put into the queue, since the causality predicate of the corresponding
action is not true.

$$\textbf{Synth-3} \qquad \frac{\beta \in \text{OUT}(S), \text{R-CAUSAL}(\beta), S_p \xrightarrow{\beta} S'_p}{S(l) \xrightarrow{\beta} S'(t(l))} \qquad\qquad (\beta = \text{out}(h(l))!)$$

This rule states that if an output action, involving a message at the head of
the queue, appears at the right-hand side of a causal relation, and if the partial
specification has a β derivation from S_p to S'_p, then S also has this derivation
and the message is taken from the queue.

$$\textbf{Synth-4} \qquad \frac{\beta \in \text{OUT}(S), \neg\text{R} - \text{CAUSAL}(\text{out}(m)), S_p \xrightarrow{\beta} S'_p}{S(l) \xrightarrow{\beta} S'(t(l))} \qquad\qquad (\beta = \text{out}(m)!)$$

This rule states that if an output action does not occur in a causal relation, then
S can execute this action if the partial specification can execute this action. The
queue is not affected.

We have chosen to memorize the messages (by placing these in the queue). Al-
ternatively, one can also put the actions themselves in the queue. Also, we do not
distinguish between output actions involving the same message; once an output
action is enabled, the corresponding message is taken from the queue. In general,
if there is not a unique relationship between messages and actions, or if the use of
messages alone is ambiguous, then one should memorize the actions themselves
(by putting them into the queue) rather than the messages. In that case rules
Synth-3 and Synth-4 need to be adapted by checking the causal relation itself.

5.3.3   The Combination Algorithm

On the basis of the above synthesis rules we are able to construct a synthesis algorithm, which we will refer to as the *COMBINATION ALGORITHM*. This algorithm can be used to synthesize the behaviour of S given a specification of the systems $S_i(0 < i < n)$ with which S has to interface. Let EXPAND be the set of states to be evaluated; initially, EXPAND contains the initial state of S (this is the n-tuple of initial states of the S_i). Let READY be the set of states that have already been evaluated. Initially, READY is the empty set.

(step 1): write the behaviours of $S_i(1 \leq i \leq n)$ in normal form.

(step 2): mirror the behaviours of the S_i to obtain the partial specification. The initial state of S is the n-tuple, formed by taking the initial state of each of the S_i.

(step 3): initially, the set EXPAND contains the initial state of S.

(step 4): define the causal relations.

(step 5): select a state from EXPAND.

(step 6): remove this state from EXPAND.

(step 7): add this state to READY.

(step 8): apply the synthesis rules for all input and output actions in the partial specification of S which are enabled (i.e. which are in $IN(S)$ or $OUT(S)$); this yields a set of behaviour equations in normal form, involving the enabled actions.

(step 9): add each next state to EXPAND whenever it is not in READY.

(step 10): when EXPAND is not empty go to step 5, else stop.

(stop): reduce the obtained behaviour of S.

In [Koo85a] it is assumed that at certain points during the application of the algorithm, the designer can interrupt the algorithm and change certain derivations that were derived by the algorithm; this is done to make the resulting behaviour of S more efficient. In the article such interventions are referred to as axioms; we will treat these axioms in Chapter 9.

5.3.4 Linking a server to a communication network

As an example of the application of the combination algorithm, let us consider the synthesis of an interface between a PABX (a private telephone exchange) and a voice mail box (*Figure 5.5*). The purpose of the service is that a telephone user calls a special number to get access to the voice mail box. The user then either records a voice message for another person, or replays a message from another person. We apply the steps of the algorithm:

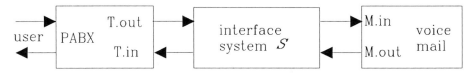

Figure 5.5: Linking a voice mail box to a telecommunication network (the user-PABX combination) by means of interface system S.

(step 1): write the behaviours of the use-PABX and the voice mail box in normal form.

The specifications of the user-PABX combination and the mailbox towards the interface system S are, respectively:

$$
\begin{aligned}
&T0 = \text{T.out!con} : T1 &&\text{a connection request is sent to } S, \text{ after}\\
&T1 = \text{T.in?ptr} : T2 &&\text{which a positive acknowledgement is}\\
&T2 = \text{T.out!control} : T3 &&\text{expected and a control signal is sent}\\
&T3 = \text{T.in?nack} : T4 &&\text{then, either a negative acknowledgement}\\
&\quad + \text{T.in?pack} : T5 &&\text{or a positive acknowledgement is expected}\\
&T4 = \text{T.out!ster} : T0 &&\text{after the session, connection is ended}\\
&T5 = \text{T.out!info} : T4 &&\text{during conversation, information is sent}\\
&\quad + \text{T.in?info} : T4 &&\text{or received, depending on the service}
\end{aligned}
$$

We assume the mailbox to have the following internal operations; a refers to the calling user (also called the A subscriber). Likewise, b refers to the called user (or B subscriber):

AUTHORIZE(op,b,a) a Boolean function, returning true or false, depending on the allowed service (specified by *op*) for the A subscriber, with respect to the B subscriber.

WRITE(a,b,info) writes the voice information *info* coming from the A subscriber to an address which will be referenced by the B subscriber for later replay.

READ(b,a) reads the voice message from the B subscriber to the A subscriber.

ERASE(b,a) erases the voice message from b when read by a.

$$M0 = \text{M.in?read(b,a)} : M1$$
$$\quad + \text{M.in?write(a,b)} : M2$$
$$M1 = \textbf{if} \text{ AUTHORIZE(read,b,a)} \textbf{ then } \text{M.out!pack} : M3$$
$$\quad + \textbf{if not} \text{ AUTHORIZE(read,b,a)} \textbf{ then } \text{M.out!nack} : M0$$
$$M2 = \textbf{if} \text{ AUTHORIZE(write,a,b)} \textbf{ then } \text{M.out!pack} : M4$$
$$\quad + \textbf{if not} \text{ AUTHORIZE(write,a,b)} \textbf{ then } \text{M.out!nack} : M0$$
$$M3 = \text{M.out!READ(b,a)} : \text{ERASE(b,a)} : M0$$
$$M4 = \text{M.in?info} : \text{WRITE(a,b,info)} : \text{CLOSE(a,b)} : M0$$

In order to get the partial specification for the interface system S, we will first simplify the equations for the mailbox:

(i) we abstract internal operations ERASE, WRITE, and CLOSE into τ's and reduce the resulting equations;

(ii) we apply the following **if-reduction rule**:

$$\frac{A = \textit{if} \text{ c } \textbf{then} \text{ B } \textbf{else} \text{ C}}{A = \tau : B + \tau : C}$$

This rule can be used if we are not interested in the details of the boolean expression c, determining the selection. Instead, we are only interested to express that A makes an internal choice. With these simplifications we get the following equations in normal form for the mailbox:

$$M0 = \text{M.in?read(b,a)} : M1 + \text{M.in?write(a,b)} : M2$$
$$M1 = \tau : \text{M.out!pack} : M3 + \tau : \text{M.out!nack} : M0$$
$$M2 = \tau : \text{M.out!pack} : M4 + \tau : \text{M.out!nack} : M0$$
$$M3 = \text{M.out!READ(b,a)} : M0$$
$$M4 = \text{M.in?info} : M0$$

(step 2): mirror the behaviours of the user-PABX and the voice mail box to obtain the partial specification. The initial state of S is the pair, consisting of the initial states of the user-PABX and mail box.

In order to perform the mirroring, we will relabel ports at the same time. We replace *T.out!* by *a?*, *T.in?* by *b!*, *M.in?* by *c!*, and *M.out!* by *d?*; in addition, we remove the τ's and obtain:

$$T0 = \text{a?con} : T1$$
$$T1 = \text{b!ptr} : T2$$
$$T2 = \text{a?control} : T3$$

T3 = b!nack : T4 + b!pack : T5
T4 = a?ster : T0
T5 = a?info : T4 + b!info : T4
M0 = c!read(b,a) : M1 + c!write(a,b) : M2
M1 = d?pack : M3 + d?nack : M0
M2 = d?pack : M4 + d?nack : M0
M3 = d?READ(b,a) : M0
M4 = c!info : M0

(step 3): the initial state of S is the pair $(T0, M0)$, written shortly as $T0M0$.

(step 4): we define the causal relations as follows:

a?control $= - >$ c!read(b,a)
a?control $= - >$ c!write(a,b)
a?info $= - >$ c!info
d?READ(b,a) $= - >$ b!info
d?pack $= - >$ b!pack
d?nack $= - >$ b!nack

We observe that the *control* message is causally related to two other messages. The designer should at this point in time know what to do with this. In the example, it is obvious that the interface system has to inspect the *control* message in order to determine whether a *read* or a *write* action is required. He has to introduce a conditional expression for this purpose. When we apply this to $M0$ we get:

M0 =   **if** control.op = read **then** c!read(b,a) : M1
       + **if** control.op = write **then** c!write(a,b) : M2

We are now ready to apply *(step 5)* - *(step 10)* of the combination algorithm. We only write the message queue if it is non empty, as was explained at the end of *Section 4.2.*

*(Synth-*2)T0M0	= a?con : T1M0
*(Synth-*4)T1M0	= b!ptr : T2M0
*(Synth-*1)T2M0	= a?control : T3M0(control)
*(Synth-*3)T3M0(control)	= **if** control.op = read **then** c!read(b,a) : T3M1
*(Synth-*3)	+ **if** control.op = write **then** c!write(a,b) : T3M2
*(Synth-*1)T3M1	= d?pack : T3M3(pack)
*(Synth-*1)	+ d?nack : T3M0(nack)
*(Synth-*1)T3M2	= d?pack : T3M4(pack)
*(Synth-*1)	+ d?nack : T3M0(nack)
*(Synth-*3)T3M3(pack)	= b!pack : T5M3
*(Synth-*1)	+ d?READ(b,a) : T3M0(pack,READ(b,a))

$(Synth\text{-}3)$T3M0(nack)   = b!nack : T4M0
$(Synth\text{-}3)$T3M4(pack)   = b!pack : T5M4
$(Synth\text{-}1)$T5M3   = d?READ(b,a) : T5M0(READ(b,a))
$(Synth\text{-}3)$T3M0(pack,READ(b,a))= b!pack : T5M0(READ(b,a))
$(Synth\text{-}2)$T4M0   = a?ster : T0M0
$(Synth\text{-}1)$T5M4   = a?info : T4M4(info)
$(Synth\text{-}3)$T5M0(READ(b,a))= b!info : T4M0
$(Synth\text{-}3)$T4M4(info)   = c!info : T4M0
$(Synth\text{-}2)$ + a?ster : T0M4(info)
$(Synth\text{-}3)$T0M4(info)   = c!info : T0M0

The choice in $T5$ between *a?info* and *b!info* is context dependent. The interface
system remembers whether a read or write access on the mailbox was intended
(as specified by the *control* message). While $T5M3$, the interface system cannot
execute *a?info* since $M3$ refers to a read access and no information is expected
from the user-PABX combination. Likewise, in $T5M4$ the message *a?info* is
enabled since $M4$ refers to a write access. In $T5M0(READ(b,a))$ the message in
the queue indicates to the interface system that a read access has been carried
out by the mailbox; the interface system therefore enables *b!info* (due to a causal
relation) but not *a?info*.

(stop): reduce the obtained behaviour of S.

The only useful simplification is the renaming of agent identifiers, including the
message list, followed by substitution, which yields (the conditional expressions
have been abbreviated to r and w respectively):

S0 = a?con : b!ptr : a?control : S1
S1 = **if** r **then** c!read(b,a) : S2 + **if** w **then** c!write(a,b) : S3
S2 = d?pack : S4 + d?nack : b!nack : a?ster : S0
S3 = d?pack : b!pack : a?info : S5 + d?nack : b!nack : a?ster : S0
S4 = b!pack : d?READ(b,a) : b!info : a?ster : S0
        + d?READ(b,a) : b!pack : b!info : a?ster : S0
S5 = c!info : a?ster : S0 + a?ster : c!info : S0

In order to verify the result, we calculate the composition of the interface system
with the user-PABX combination. Hence, we calculate $C(S0, T0)$, and after
applying *if*-reduction, we find:

S0T0 = τ : S1T3
S1T3 = τ : c!read(b,a) : S2T3 + τ : c!write(a,b) : S3T3
S2T3 = d?pack : d?READ(b,a) : S0T0 + d?nack : S0T0
S3T3 = d?pack : c!info : S0T0 + d? nack S0T0

Exercise 5.3. Show that $C(S0, T0) \approx^m M0$. ■

Exercise 5.4. Calculate $C(S0, M0)$ and develop a similar reasoning as for $C(S0, T0)$. ■

Exercise 5.5. Consider the diagram of *Figure 5.6*; the equations for agents A, B and the user terminal T are:

```
A  = a? : A1 + b? : A2    B =  c? : B1 + d! : B2    T =  bus2!execute : T
A1 = c! : A3              B1=  e? : B                   + bus2!request : T1
A2 = d? : A3              B2=  y! : B1               T1 = bus2?answer : T
A3 = e! : A
```

   (a) Calculate $C(A, B)$;

   (b) Determine the program for the microcomputer; the following causal relations are given:
       $bus2?execute =-> a!; \quad bus2?request =-> b!; \quad y? =-> bus2!answer;$

   (c) Synthesize the interface between the microcomputer and the user. ■

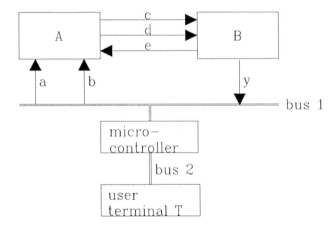

Figure 5.6: A microcontroller as an interface system, linking a bus with two circuits with a user terminal.

5.4   Completing the specification

In case we have a specification SPEC, and an agent A which implements SPEC partially, how can we calculate the missing part S, such that $C(A, S) = $ SPEC? Without further proof, it can be shown that $\overline{S} \subseteq C(A, \overline{SPEC})$.

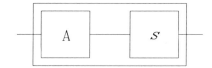

Figure 5.7: Given a specification SPEC and a partial implementation A, find the missing part S such that $C(A, S) = $ SPEC.

Let us use our well known handshake system to try this out (see *Section 2.6*). We assume we have the specification C and the agent A. The question is, what is the required behaviour for agent B?

$$
\begin{aligned}
A &= \text{a.in1? : A1} & C &= \text{a.in1? : C1} \\
A1 &= \text{a.out! : A2} & C1 &= \text{b.out1! : C} \\
A2 &= \text{a.in2? : A}
\end{aligned}
$$

As a first step, we write $M(C)$, the mirror image of C:

$$
\begin{aligned}
\text{c} = \text{a.in1! : c1} \qquad \text{c1} = \text{b.out1? : c}
\end{aligned}
$$

Expansion yields:

$$
\begin{aligned}
\text{Ac}   &= \tau : \text{A1c1} \\
\text{A1c1} &= \text{a.out! : A2c1 + b.out1? : A1c} \\
\text{A2c1} &= \text{a.in2? : Ac1 + b.out1? : A2c} \\
\text{A1c} &= \text{a.out! : A2c} \\
\text{Ac1} &= \text{b.out1? : Ac} \\
\text{A2c} &= \text{a.in2? : Ac}
\end{aligned}
$$

We have to take the mirror image of the above equations; hence, we replace *a.out!* by *b.in?*, *b.out1?* by *b.out1!*, and *a.in2?* by *b.out2!* which, after substitution, and renaming agent identifiers, yields:

$$
\begin{aligned}
\text{B} &= \tau : \text{B0} \\
\text{B0} &= \text{b.in? : B1 + b.out1! : B2} \\
\text{B1} &= \text{b.out2! : B3 + b.out1! : B4} \\
\text{B2} &= \text{b.in? : B4} \\
\text{B3} &= \text{b.out1! : B} \\
\text{B4} &= \text{b.out2! : B}
\end{aligned}
$$

Exercise 5.6. Show that $C(A, B)$ yields:

> AB = τ : (a.in1? : X + b.out1! : a.in1? : AB)
> X   = a.in1? : b.out1! : X + b.out1! : AB

∎

From the expression for $C(A, B)$ we observe that the following action sequences could occur repeatedly:

> τ : a.in1? : a.in1? : b.out1!
> τ : a.in1? : b.out1!
> τ : b.out1! : a.in1?

Obviously, we are only interested in the second sequence. By filtering the sequences such that only those allowed by the specification C would be produced, we are able to adapt the equations for B. From the expression for B we observe that three action sequences are possible:

> τ : b.in? : b.out1! : b.out2!
> τ : b.in? : b.out2! : b.out1!
> τ : b.out1! : b.in? : b.out2!

Only the first sequence is the one we would like to have. Therefore, like in the case of interface derivation of *Section 5.3*, we have to put constraints on the possible sequences of B. If we apply the following causal relation: *a.in1?* $= ->$ *b.out1!*, then the third sequence would be removed. In that case the behaviour for B would be:

> B = τ : b.in? : (b.out2! : b.out1! : B + b.out1! : b.out2! : B)

Adding a second causal relation *b.out1!* $= ->$ *b.out2!*, then the equations for B would become:

> B = τ : b.in? : b.out1! : b.out2! : B

Exercise 5.7. Show that with the new equations for B we find that $C(A, B) \approx C$.

∎

Hence, we would not satisfy $C(A, B) = C$. However, the observation equivalence is sufficient if we do not use C in another context. Again, we can decide that mirror observation equivalence would be sufficient in this case. In conclusion, *completing the specification* is a technique which can be applied by screening the resulting expressions for the allowed action sequences as prescribed in the specification.

Chapter 6

CCS AND SDL

6.1 A short overview of SDL

The CCITT Specification and Description Language SDL [CCI88a, SPT87] is intended as a language to write specifications for telecommunication applications, such as telephone exchange control systems. It is widely used within the telecommunications industry. In other areas, such as in process control systems, SDL may find additional applications. *Figure 6.1* illustrates the major SDL concepts. An SDL specification consists of *states* and *transitions*. Transitions consist of *communication actions*, *decision points* and *tasks*. These elements are connected by means of *flow lines*; *connectors* are used to link SDL diagrams together.

A **state** is a condition in which the action of a process is suspended awaiting an input. A **transition** is a sequence of actions which occurs when a process changes from one state to another in response to an input. A process can be either in one of its states or in a transition at any one instant. An **input** is an incoming signal which is recognized by a process. An **output** is an action within a transition which generates a signal which in turn acts as an input elsewhere. A **save** is the postponement of recognition of a signal when a process is in a state in which recognition of that signal does not occur. A **task** is used in a transition to represent a set of operators on variables. A **decision** consists of an expression and a number of paths leaving the decision; each value of the expression correspondents with one such path.

A state can be followed by inputs or saves only; inputs can be followed by tasks, outputs and decisions; saves are not followed by any symbol. With respect to signal reception the following conventions hold. When a signal arrives at a process, it is considered to be retained for that process (i.e. it is not yet consumed by it). This means that there is an implicit queuing mechanism. Retained signals are stored in a first-in-first-out order. This provides the possibility to apply the QCom rules of *Section 4.1*. Saved signals (i.e. signals input via saves) are stored for use in later transitions. The save concept is somewhat peculiar and is a consequence of the FIFO-order in storing signals.

93

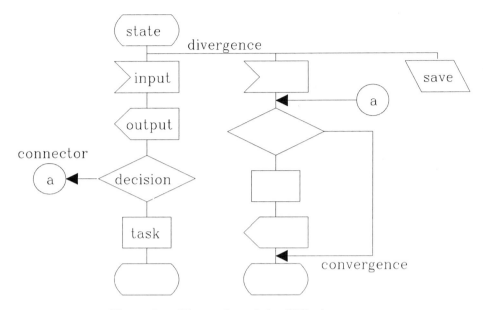

Figure 6.1: Illustration of the SDL elements.

In SDL diagrams a sequence of output symbols may follow an input symbol. In the SDL report there is no constraint put on the order in which signals are sent. In [Shi82] it is therefore assumed that signals are sequenced according to their output symbols. Accordingly, the class of SDL systems which satisfy this rule are indicated as Unambiguous SDL-systems (USDL-systems). For this type of systems it is proved that queue overflow and deadlock are undecidable, i.e. there are no algorithms for determining whether queue overflow or deadlock will occur for an arbitrarily given USDL-system.

6.2   Linking CCS and SDL

In SDL, states must be followed by inputs. Such a rule is absent in CCS. *Agent identifiers* in CCS correspond to *states* and *connectors* in SDL. To simplify matters, we will only use the connector symbol (*Figure 6.2*). During input or output, transfer of information between two processes, each represented by an SDL diagram, takes place. The implicit queuing mechanism can be accounted for by means of the method presented in *Section 4.2*. Decision points can be accounted for by using conditional behaviour expressions. Another possibility would be to apply *if-reduction* using the silent action τ (*Figure 6.3*). *Divergence* is accounted for in CCS by the *summation operator*. *Convergence* means that an agent identifier can occur in one or more other behaviour expressions (in analogy: a state may be reachable from more than one other state).

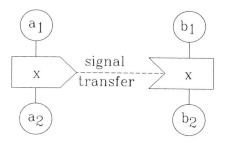

Figure 6.2: Illustration of input/output, with $out! = \overline{in?}$.

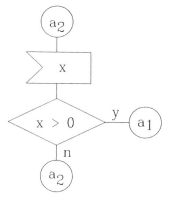

Figure 6.3: Decision point in SDL, and its CCS representation using a conditional expression, or using *if-reduction*.

Tasks are a different matter. As long as tasks only have a local effect (i.e. there are no side-effects), then tasks can be interpreted as τ *actions*. In case tasks have side-effects, then interpreting these tasks becomes more complicated as these side-effects have to be precisely modeled in CCS. Another feature which is not defined in CCS, but which might be helpful is the use of programmed labels, in which names of output ports are calculated before doing the output action; this is used in [OP90] for describing protocols in mobile communication systems.

In conclusion, relating CCS and SDL is possible under certain restrictions. As was done in [Shi82], the ordering of output signals should be in correspondence with the ordering in which the associated output symbols appear in a transition (i.e we consider USDL systems only). In addition, the input of signals should obey the FIFO rule. Relations between SDL and other formalisms can be found in [Gel87] and [Hog89].

6.3   A pipeline

Figure 6.4a shows the SDL diagram of two communicating agents.

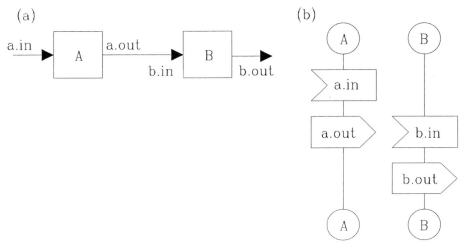

Figure 6.4: (a) Systems a and *b* and (b) the SDL diagrams of their behaviour.

The agents *a* and *b* are specified by the following set of equations:

        Agent a::   A = a.in?x : a.out!x : A.
        Agent b::   B = b.in?x : b.out!x : B.

6.4   A handshake system

If SDL diagrams can be put in terms of CCS, we can also apply CCS laws to
these diagrams. Our handshake system is shown in *Figure 6.5*. From *Section
2.4* we know the behaviours of *A* and *B* in terms of CCS while in *Exercise 3.1*
we have shown the combined behaviour of A and B to be observation congruent
with the behaviour of agent $C = (A \mid B)\backslash\{a.out, a.in2, b.in, b.out2\} = a.in1?$:
$b.out1!$: C.

6.5   A network protocol

We will consider the protocol developed in [Koo85a], which we will treat extensively in *Chapter 9*. The protocol prescribes the communication between a
telecommunication network and subscribers. *G.in1* and *G.out1* denote the input and output ports of the network, connected to the A-subscriber (i.e. the
subscriber who initiates a call). Likewise, *G.in2* and *G.out2* denote the input
and output ports of the network, connected to the *B*-subscriber (the subscriber

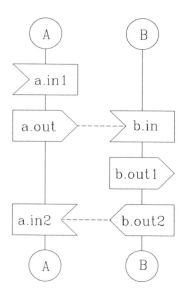

Figure 6.5: SDL diagram of the handshake mechanism.

receiving the call). The messages *con, ster, ptr, ntr* and *dter* are described in *Section 9.1. Figure 6.6* shows the SDL diagram of the equations derived in *Sections 9.2* and *9.3.*

6.6   SS7 call processing control

As a last example, we will consider the SS7 call processing control as described in [CCI88b]. *Figure 6.7* shows the SDL diagram (it is the first diagram out of the seven diagrams describing the full protocol). The SDL specification can be translated into CCS. After applying *if-reduction* for the conditions, this yields the following set of equations:

$$
\begin{aligned}
\text{Idle} =\ & \text{IAM? : S1} \\
& + \text{CLF? : RLG! : Idle} \\
& + \text{RLG! : Idle} \\
& + \text{other message: start\_CPC-CCI! : Idle} \\
& + \text{RSC? : S2} \\
\text{S1} =\ & \tau : \text{S3} + \tau : \text{S4} \\
\text{S2} =\ & \tau : \text{S10} + \tau : \text{S11} \\
\text{S3} =\ & \tau : \text{S4} + \tau : \text{S5} \\
\text{S4} =\ & \text{seize : S6} \\
\text{S5} =\ & \text{send\_BLO! : Idle} \\
\text{S6} =\ & \tau : \text{S7} + \tau : \text{S8} + \tau : \text{S9}
\end{aligned}
$$

$S7 =$ start_T1! : S8
$S8 =$ start_OGC! : WFR
$S9 =$ start_CPC-CCI : S7
$S10 =$ unblock! : S11
$S11 = \tau : S12 + \tau : S13$
$S12 =$ send_BLO! : RLG! : Idle
$S13 =$ RLG! : Idle.

(WFR not further specified here).

$S1$ and $S3$ can be reduced using the τ-4 law. Substituting $\tau : S4$ for $S1$ and applying the τ-1 law to *Idle* yields:

Idle =   IAM? : S4
       + etc.

SS7 uses three types of messages;

(a) messages to/from other telephone exchanges;

(b) messages within the call processing control part, and

(c) messages to/from other parts outside the call processing control, but within the same exchange (e.g. messages to the signaling procedure control).

One can use the translation into CCS in order to analyse the SS7 protocol. For instance, if one is interested in the behaviour of the protocol *between* exchanges, one could apply the following simplifications

(step 1) replace all internal communications by τ's;

(step 2) replace each task by a τ;

(step 3) apply τ-laws.

(step 5) to simplify further, messages can be stated in terms of the messages which are used for the telecommunication protocols as derived in *Chapter 9*.

Here are some examples:

IAM	con1	(the connection request
SAM	con2	is split up into several messages)
ACM	pack(con)	(a positive acknowledgement on receipt of the connection request)
CLF	ster	(source termination request)
CBK	dter	(destination termination request)
COT	pdr	(positive data response: data has been received successfully)
ANC	ptr	(positive termination response: a connection has been established)
RLG	pack(ster)	(a positive acknowledgement on receipt of the source termination request)

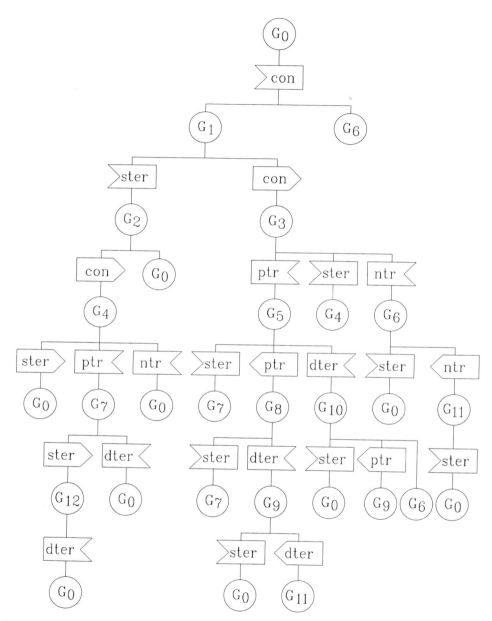

Figure 6.6: SDL diagram of the specification of the network protocol (the communication behaviour of the network) with respect to the A- and B-subscribers. The names of the sending and receiving ports of control messages have been omitted.

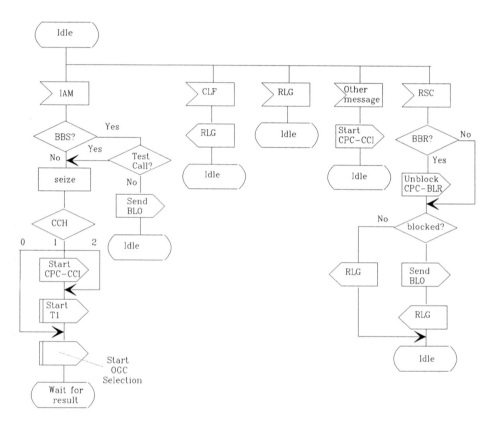

Figure 6.7: First SDL diagram of the SS7 call processing control protocol.

Chapter 7

CCS AND PETRI NETS

7.1 From CCS to Petri Nets

It will be assumed that the reader is familiar with the notion of a Petri Net. For an introduction to Petri Net Theory, see [Pet81] and [Bra80]. We will consider *agent identifiers, actions, summation, restriction,* and *parallel composition* and relate these CCS concepts to the corresponding Petri Net primitives. It can be shown that for each behaviour tree, a corresponding Petri Net can be found. This illustrates that behaviour tree structures in CCS are a proper subset of the class of Petri Nets. More in particular, CCS trees correspond to 1-safe Petri Nets. A **safe Petri Net** satisfies the condition that no place will contain more than one token. This constraint assures that the net is bounded.

Agent identifiers correspond to the *places* of a Petri Net of the same system. *Actions* correspond to the *transitions* in the Petri Net. As branches in a behaviour tree always point from a node (agent identifier) to another node, this means that with each action in the behaviour tree there corresponds a transition with exactly one input place and one output place; this is illustrated in *Figure 7.1.*

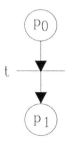

Figure 7.1: A Petri Net for the CCS equation $p_0 = t : p_1$.

The relationship as shown in *Figure 7.1* is a static one; the dynamic case is as follows: whenever control reaches agent identifier p_0 in the behaviour equation, a token is generated in place p_0. If the action has taken place (i.e. if the transition fires) then the token is put into the output place.

In the case of *summation* an agent can select one of several actions. Within Petri Nets this situation corresponds to *choice*, where a place has multiple output transitions.

Restriction means removing certain actions as well as the resulting subtrees from the behaviour tree. In terms of Petri Nets this means that the corresponding transitions have to be removed; the same applies for any resulting disconnected nets (a net $PN2$ is disconnected from another net $PN1$ if there is no transition which leads from $PN1$ to $PN2$).

The case of *parallel composition* goes as follows. If we wish to calculate the behaviour of a composite system by means of the expansion law, then we have to define which ports communicate with each other. In terms of Petri Nets this means that the Petri Nets of the subsystems communicate via one or more transitions: each of these transitions has multiple inputs and outputs; in the case of A and B in the handshake example of *Section 2.6* a transition representing communication between A and B has two inputs and two outputs. *Figure 7.2a* shows the Petri Nets of systems A and B, while *Figure 7.2b* shows their combined Petri Net. If we reduce the corresponding CCS expressions using the τ laws, then we obtain the Petri Net of *Figure 7.2c*; the dotted lines in *Figure 7.2b* show the abstraction involved in this reduction. Hence, putting systems into communication means identifying transitions in the subnets.

The translation of unparameterized CCS equations without value-passing into a Petri Net yields a safe net. This can be seen as follows. With each agent identifier a place is associated in the corresponding Petri Net. When control reaches a particular agent identifier, this corresponds to the generation of a token in the place associated with that agent identifier. Hence, with a flow of control through a behaviour tree one can associate a corresponding flow of control in the Petri Net involving a single token. Another way of putting this is by saying that the net is safe since we do not employ parameterized agent identifiers. For each CCS equation (or equivalently: for each behaviour tree) the resulting net contains a single token and is conservative (a Petri Net is conservative if the number of tokens is constant).

7.2   From Petri Nets to CCS

For the reverse case (from Petri Nets to CCS) we have to consider the general situation in which a transition has n input places and m output places *(Figure 7.3a)*. In order to analyze the relation with CCS for this general case, we will consider two extremes; one in which we have a transition with one input place and multiple output places *(Figure 7.4)*, and one in which we have a transition

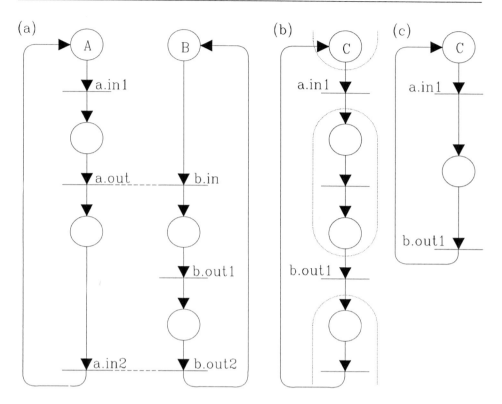

Figure 7.2: *(a)* Petri nets of the two subsystems *A* and *B*; *(b)* systems *A* and *B* put together to communicate and *(c)* the resulting Petri Net of the handshake system after reduction.

with multiple input places and one output place *(Figure 7.5)*. The relation with the general case is then straightforward.

If we have multiple output places, then in terms of CCS this means that agent p_0 triggers the parallel composition of multiple agents. *Figure 7.4* shows the static case. The dynamic case follows again from the observation that whenever control reaches agent identifier p_0, a token is generated in place p_0(and vice versa).

Obtaining the CCS expression for the case with multiple input places (*Figure 7.5*) is straightforward by reversing the direction of the transition. In terms of CCS this means that we have to reverse the roles of the agent identifiers at both sides of the equal sign. The resulting equation is given in *Figure 7.5*; as the transition can only fire whenever there is a token in all its input places, in terms of CCS this means that several agents are put into parallel composition, from which only one action is possible. In fact we have to interpret the parallel composition of these agents as a single new agent identifier.

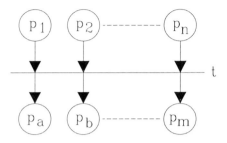

Figure 7.3: A transition with n input places and m output places; the corresponding CCS equation is: $(p_1 \mid p_2 \mid \ldots \mid p_n) = t : (p_a \mid p_b \mid \ldots \mid p_m)$.

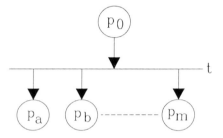

Figure 7.4: A transition with one input place and several output places; the corresponding CCS equation is : $p_0 = t : (p_a \mid p_b \mid \ldots \mid p_m)$.

To obtain the result for the general case of *Figure 7.3* we have to combine the results of the cases shown in *Figures 7.4 and 7.5*. We do this by identifying place p_0 in *Figure 7.4* with place p_0 in *Figure 7.5* and abstracting the Petri Net made up of transition t (from *Figure 7.5*), place p_0 and transition t (from *Figure 7.4*) into one single transition t. Hence, we abstract a Petri Net which starts at a transition and ends in a transition into one single transition t. We obtain the following CCS equation:

$$(p_1 \mid p_2 \mid \ldots \mid p_n) = t : (p_a \mid p_b \mid \ldots \mid p_m) .$$

Hence, we have to interpret a transition with multiple inputs and multiple outputs as an action, whereby the parallel composition of some agents produces the parallel composition of some other agents.

7.3   Communication via a shared buffer

Figure 7.6 shows the Petri Net of two processes, a producer process and a consumer process. These processes communicate via a shared buffer (not shown).

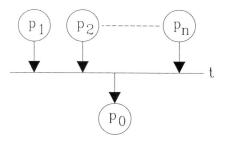

Figure 7.5: A transition with m input places and one output place; the corresponding CCS equation is: $(p_1 \mid p_2 \mid \ldots \mid p_n) = t : p_0$.

Mutually exclusive access to the buffer is controlled by two semaphores *empty* and *full*. The producer process produces a message, performs a P operation on semaphore *empty*. If this succeeds, it puts a message in the buffer. Next, it performs a V operation on the semaphore *full*. The consumer process then performs a P operation on *full*. If this is done, it has access to the buffer and it can take the message out of the buffer. The consumer process then performs a V operation on *empty* and consumes the message, after which the whole cycle can start again.

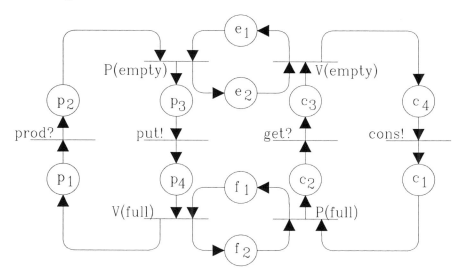

Figure 7.6: Petri net of a producer/consumer system, where two processes (the producer and consumer) communicate via a shared buffer, the access to which is determined by two semaphores *empty* and *full*.

We will model the P and V operations as inputs and outputs. As an example,

P(empty)? denotes an input action via the (fixed) port *P(empty)*; it models the
P operation on the semaphore *empty*. Likewise, the production of a message
by the producer will be modeled as the receipt of that message, whereas the
consumption of that message by the consumer will be modeled as the sending
of the message. We will assume an initial marking of the Petri Net to consist
of tokens in places $p1$, $e1$, $f1$ and $c1$. For this marking, the CCS equations
associated with the producer, consumer and semaphores are:

$$
\begin{array}{llll}
\text{prod::} & p1 = \text{prod?} & :p2 \\
& p2 = \text{P(empty)?} & :p3 \\
& p3 = \text{put!} & :p4 \\
& p4 = \text{V(full)!} & :p1 \\
\text{empty::} & e1 = \text{P(empty)?} & :e2 \\
& e2 = \text{V(empty)!} & :e1 \\
\end{array}
\qquad
\begin{array}{llll}
\text{cons::} & c1 = \text{P(full)?} & :c2 \\
& c2 = \text{get?} & :c3 \\
& c3 = \text{V(empty)!} & :c4 \\
& c4 = \text{cons!} & :c1 \\
\text{full::} & f1 = \text{V(full)!} & :f2 \\
& f2 = \text{P(full)?} & :f1 \\
\end{array}
$$

We can calculate the combined behaviour of these processes, and hence the behaviour of the Petri Net as a whole, by performing a parallel composition on
these equations:

$$\text{prodcon} = (p1 \mid c1 \mid e1 \mid f1)\backslash\{\text{P(empty)}, \text{V(empty)}, \text{P(full)}, \text{V(full)}\}$$

Performing the expansion and reducing the equations yields the following behaviour equations (*Figure 7.7* shows the corresponding behaviour tree):

$$
\begin{array}{ll}
\text{prodcon::} & m1 = \text{prod?: put!: } m2 \\
& m2 = \text{prod?: get?: } m3 + \text{get?: } m4 \\
& m3 = \text{put!: } m5 + \text{cons!: put!: } m2 \\
& m4 = \text{prod?: } m3 + \text{cons!: } m1 \\
& m5 = \text{prod?: cons!: get?: } m3 + \text{cons!: } m2 \\
\end{array}
$$

Exercise 7.1. Try a particular firing sequence of the Petri Net and determine the
corresponding path in the behaviour tree.   ∎

We can also hide the actions *prod?* and *cons!* by using two dummy processes X
and Y ($X = prod! : X$ and $Y = cons? : Y$). However, the same effect can be
obtained by relabelling:

$$\text{prodcon}' = \text{prodcon}[\tau/\text{prod?}, \tau/\text{cons!}]$$

we find:

$$\text{prodcon}' = \tau : \text{put!} : \text{get?} : \text{prodcon}'$$

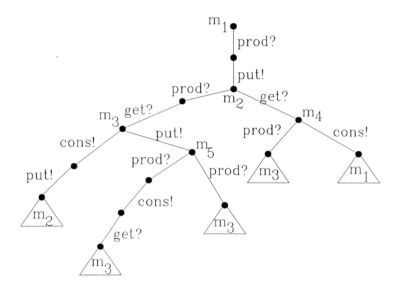

Figure 7.7: Behaviour tree of the producer/consumer system.

as we would expect (for observation equivalence we can also remove the τ). We observe that the access to the buffer is mutually exclusive; after a *put!* action only a *get?* action is possible.

7.4 Communication between a printer and a printer monitor

The previous section considered the translation of a Petri Net into CCS. In this section we will consider the other direction. As an example we will use a system comprising a printer and a printer monitor. The equations for the printer monitor PM and printer P are:

$$
\begin{aligned}
PM_0 &= \text{pm.in1?prt} : PM_1 \\
PM_1 &= \tau : PM_2 + \tau : PM_3 \\
PM_2 &= \text{pm.out2!eof} : PM_0 \\
PM_3 &= \text{pm.out2!prt} : \text{WFP} \\
\text{WFP} &= \text{pm.in2?ready} : \text{WFP'} \\
\text{WFP'} &= \text{pm.out1!ready} : PM_0 \\
\text{PRT} &= \text{p.in?prt} : \text{PRT1} + \text{p.in?eof} : \text{PRT} \\
\text{PRT1} &= \text{p.out!ready} : \text{PRT}
\end{aligned}
$$

Figure 7.8 shows the block diagram of the printer and the monitor. The Petri Net corresponding to the above behaviour equations is shown in *Figure 7.9 (a)-(b)*. Calculating the combined behaviour of printer and monitor requires their

Petri Nets to be connected at the transitions corresponding to their mutual communication actions, as shown in *Figure 7.9 (c)*.

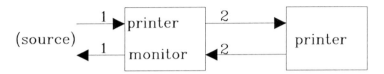

Figure 7.8: Block diagram of printer and printer monitor.

Calculating $C(PM_0, PRT)$ and reducing the resulting equations, yields:

X0 = pm.in1?prt : X1
X1 = τ : X0 + τ : pm.out1!ready : X0

The resulting Petri Net is shown in *Figure 7.9 (d)*. The transformation of the Petri Net of *Figure 7.9 (c)* into that of *Figure 7.9 (c)* yields a *net morphism*.

7.5   An assignment statement

So far we have only dealt with pure synchronization. If we are interested in value-passing as well, we have to consider the question as to how values are represented in both CCS and Petri Nets. CCS agents have input ports, output ports and local variables. In the case of output actions, values are assigned to output ports, while for input actions values received via input ports are assigned to local variables. As an example of value passing we will show how an assignment statement can be implemented using communication between two agents; one of them producing a value and the other storing this value. For instance, consider the two communicating systems X and Y defined as follows:

X = trig? : port!x : X   and   Y = port?y : Y

Both systems communicate via *port*. The triggering input *trig* starts the communication. Putting X and Y together, we get the combined behaviour:

B = (X|Y)\port = trig? : τ : B(x/y),

where $B(x/y)$ is the combined behaviour of X and Y with the value of y replaced by x everywhere (in this case at one location). Agent B implicitly specifies an assignment statement. We can hide the triggering input by defining a trigger process as follows:

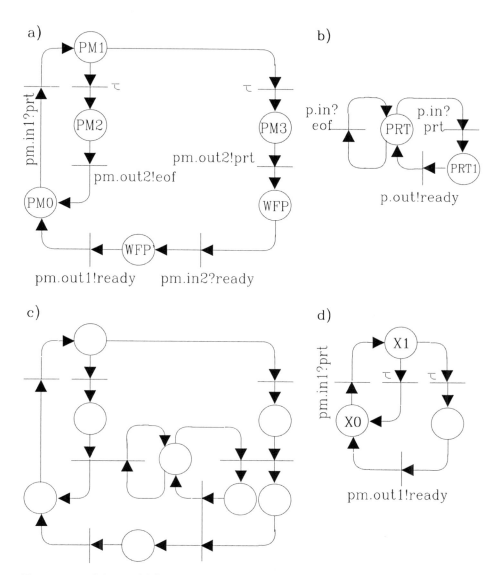

Figure 7.9: *(a)* and *((b))* Petri Nets of printer monitor and printer; *(c)* Petri Net of their combination and *(d)* Petri Net of the combination after reduction.

TRIG = trig! : NIL

Putting the trigger process in communication with X and calculating the combined behaviour of X, Y and TRIG yields:

(X|Y|TRIG)\\{port,trig}
= τ : τ : (X|Y|NIL)\\{port,trig}(x/y)
= τ : τ : (X|Y)\\{port}(x/y).
= τ : (X|Y)\\{port}(x/y).
$\approx$ (X|Y)\\{port}(x/y).

Within Petri Nets we can only represent values in terms of sets of tokens; this implies that we cannot use variables but we have to use specific values. In fact, we have to restrict ourselves to non-negative integers; a particular integer value n corresponds to a place holding n tokens.

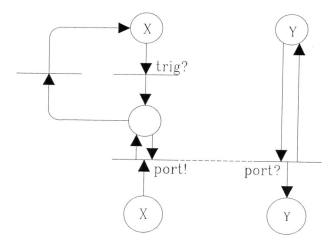

Figure 7.10: Petri Net of a triggered assignment by means of communication.

According to *theorem 2.1* [JV80] (page 181), Petri Nets can be used to weakly compute polynomials with non-negative integer coefficients. For instance, calculating the function $y := x$ can only be done in the weak sense, which means that after performing the associated computation in the corresponding Petri Net, y will have a value between 0 and $x : 0 \leq y \leq x$. *Figure 7.10* shows the Petri Net for weakly computing the previous function, where x is chosen to have the value five. In the next example it will be shown how a function can be computed in the normal (non-weak) sense.

7.6 A multiplier

The previous simple example illustrated the conversion of value-passing in CCS into Petri Net form. A somewhat larger example will be used for the reverse case. Consider the Petri Net of *Figure 7.11*. It represents a multiplier, of which the required mode of working is as follows (how this mode of working is imposed on the net will be explained later). PX and PY are the places holding the two argument values x and y. Place CNT will contain the result $x \times y$. Initially the multiplier is in a state in which place $P1$ contains a token. Now suppose PX contains three tokens, whereas PY contains two tokens. We see that only transition $t1$ is enabled; its effect is that one token is removed from PX. Now the idea is to generate as many tokens in CNT as there are tokens in PY for each token in PX. The token in $P2$ will generate y tokens in CNT as well as in place $P3$. When all tokens of PY have been copied into CNT and $P3$, the multiplier will put the tokens in $P3$ back into PY by first firing transition $t2$ and then copying $P3$ into PY. Then $t1$ is fired again, after which the next copy of y is made into CNT and $P3$ until no tokens are left in PX and y tokens are left in PY. We can see that the value of x is consumed (destructive reading) whereas the value of y is restored (we could have chosen to restore the value of x as well using a similar mechanism as for PY).

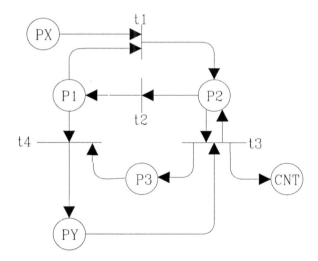

Figure 7.11: Petri Net of a multiplier.

Because of the weak computation property, we have to impose certain constraints in order to have the required multiplication. This required mode of working is imposed as follows. We see that the net contains two places with choice, $P1$ and $P2$. For instance, whenever the places $P1$ and $P3$ contain a token, both transitions $t1$ and $t4$ are enabled. By assigning *priorities* to the arcs leading to these transitions, we can impose a certain constraint on the order in which the transitions can fire. For example we can assign the priority 1 to the arc leading to $t4$ and a priority 0 to the arc leading to $t1$. Assuming that 1 is the highest possible value and 0 the lowest possible value, this would mean that in case of a conflict it is always $t4$ that is fired. If there is no conflict, as in the case where $P3$ has no token, $t1$ can be fired. Instead of speaking in terms of probabilities, 0 and 1 are the probabilities that $t4$ or $t1$ can occur; a 0 value then means the impossible case, whereas the value 1 means the certain case. In a similar fashion, we assign priorities 0 and 1 to the arcs leading to transitions $t2$ and $t3$ respectively.

In terms of CCS, we want the behaviour of the multiplier to be:

$$\text{MULT}(3, 2, 0) = \tau\colon \text{MULT}(0,2,6).$$

We can take this expression as the specification of the behaviour of the net; the τ indicates that after some internal action (the multiplication) the resulting behaviour is MULT(0,2,6). For observation equivalence the τ can be removed, and we get:

$$\text{MULT}(3,2,0) \approx \text{MULT}(0,2,6).$$

In order to verify that this is indeed the case, we split the Petri Net into three components (*Figure 7.12*). Let $p(n)$ denote the place p with n tokens. Then using the marking $\{PX(x), P1(1), PY(y)\}$, the behaviour of the three subnets is as follows. For the network of *Figure 7.12(1)* we derive:

$$\text{P1} = (\text{P1} \mid \text{CNT}(0)).$$

CNT(0) means place *CNT* with zero tokens. In terms of CCS this corresponds to *NIL* behaviour. Because $P1 = (P1 \mid NIL)$ (one of the CCS laws), we may replace *NIL* by *CNT(0)*. With priorities written between round brackets, the previous expression can be rewritten as:

P1 | CNT(0) = (0) t1! : (P2 | CNT(0)) + (1) t4! : (P1 | CNT(0)).
P2 | CNT(0) = (0) t2! : (P1 | CNT(0)) + (1) t3! : (P2 | CNT(1)).
P2 | CNT(1) = (0) t2! : (P1 | CNT(1)) + (1) t3! : (P2 | CNT(2)).

etc...

We can generalize this using the following parameterization:

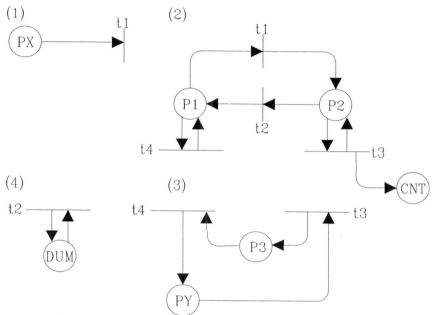

Figure 7.12: Splitting the Petri Net of *Figure 7.11* into three constituents (1), (2) and (3). The dummy process (4) is needed to hide *t2* in the parallel composition as shown in *appendix A*.

P1 | CNT(c) = (0) t1! : (P2 | CNT(c)) + (1) t4! : (P1 | CNT(c)).
P2 | CNT(c) = (0) t2! : (P1 | CNT(c)) + (1) t3! : (P2 | CNT(c+1)).

where c is a local variable, of which the scope is the corresponding expression. Likewise, for the subnets of *Figure 7.12(1)* and *Figure 7.12(3)* we get:

PX(x)   = t1! : PX(x-1)(x > 1)
PX(1)   = t1! : NIL.
PY(y)   = PY(y) | P3(0) (notice that P3(0) has NIL behaviour)
                = t3! : (PY(y-1) | P3(1))
PY(y) | P3(com) = t3! : (PY(y-1) | P3(com+1)) + t4! :
                  (PY(y+1) | P3(com-1)).
PY(0) | P3(com) = t4! : (PY(1) | P3(com-1)), with com>0.

For the multiplier we get:

MULT(x,y,0) = (P1 | CNT(0) | PX(x) | PY(y) | P3(0))\{t1,t3,t4}

In *appendix A* it is shown that this composition has the required behaviour as indicated in the expression for the multiplier.

Part II

A DESIGN METHOD FOR COMMUNICATING SYSTEMS

Chapter 8

DESIGN PROCESS DESCRIPTIONS

8.1 A programming example

In his book *The Science of Programming*, Gries [Gri84] develops a formal basis for program development, making explicit the steps and strategies associated with it. In view of our desire to formalize the design process by building suitable models of its underlying mechanisms and structures, it seems very worthwhile to consider an example from the above book. Before considering this example, we need the following notation:

$\{Q\}S\{R\}$ with Q and R predicates, and S a program,

has the following interpretation: "if execution of S is begun in a state satisfying Q, then it is guaranteed to terminate in a finite amount of time in a state satisfying R". Q is the **precondition** and R the **post-condition** of S. In fact, the expression $\{Q\}S\{R\}$ is itself a predicate, since throughout program development it is our business to assure that $\{Q\}S\{R\}$ remains true. The weakest precondition $wp(S, R)$ is the set of all states such that execution of S begun in any one of them is guaranteed to terminate in a finite amount of time in a state satisfying R.

We say that b is weaker *than c if $c \Leftrightarrow b$; the predicate wp is called weakest precondition since for any Q satisfying $\{Q\}S\{R\}$ the following implication holds: $Q \Leftrightarrow wp(S, R)$.*

Let us consider the derivation of a program using an alternative construct. The example is from [Gri84] (p.172) and italics are used to show the quotations from Gries' book (T stands for *true*):

"Write a program that, given fixed integers x y, sets z to the maximum of x and y. (...) Thus, a command S is desired that satisfies:

(14.1) { T} S { R: z = max (x,y) }.

Before the program can be developed, R must be refined by replacing max by its definition - after all, without knowing what max means one cannot write the program. Variable z contains the maximum of x y if it satisfies

(14.2) R: $z \geq x \wedge z \geq y \wedge (z = x \vee z = y)$.

Now, what command could possibly be executed in order to establish (14.2)? Since (14.2) contains z = x, the assignment z := x is a possibility".

In fact, Gries makes use of some properties of the alternative construct and applies the following strategy from [Gri84] (p.174):

"(14.7). STRATEGY FOR DEVELOPING AN ALTERNATIVE COMMAND: to invent a guarded command, find a command C whose execution will establish post-condition R in at least some cases; find a Boolean B satisfying B ⇔ wp(C, R); and put them together to form B → C(...). Continue to invent guarded commands until the precondition of the construct implies that at least one guard is true (...)".

Let us proceed with the derivation of the algorithm:

"To determine the conditions under which execution of z := x will actually establish (14 .2), simply calculate wp("z := x",R):

$$wp ("z := x", R) \quad = x \geq x \wedge x \geq y \quad \wedge (x = x \vee x = y)$$
$$= T \wedge x \geq y \qquad \wedge (T \vee x = y)$$
$$= x \geq y$$

This gives us the conditions under which execution of z := x will establish R, and our first attempt at a program can be:

if $x \geq y \rightarrow z := x$ **fi**

This program performs the desired task provided it doesn't abort. Recall from Theorem 10.5 for the Alternative Construct that, to prevent abortion, precondition Q of the construct must imply the disjunction of the guards, i.e. at least one guard must be true in any initial states defined by Q. But Q, which is T, does not imply $x \geq y$. Hence, at least one more guarded command is needed.

Another possible way to establish R is to execute z := y. From the above discussion it should be obvious that $y \geq x$ is the desired guard. Adding this guarded command yields

(14.3) **if** $x \geq y \rightarrow z := x$
        $[\![y \geq x \rightarrow z := y$
         fi

Now, at least one guard is always true, so that this is the required program".

We can make the following observations:

- it was decided that the program should have the form of an alternative construct;

- the program was developed from a specification, given by the precondition T and the post condition R (and some textual explanation);

- the program development proceeded in a number of development steps;

- use was made of certain knowledge and properties of the alternative construct.

- a verification of the above program is done using a certain theorem which we will not repeat here.

Gries also treats another example, the derivation of a program based on an iterative construct [Gri84] (p.195). Both exercises have a similar overall structure: a decision is taken on the strategy for the implementation. Then, an implementation is derived from a specification using certain knowledge about the constructs, followed by verification. Such examples illustrate nicely some notions we want to use to describe design processes, not only of algorithms, but of software and hardware in general. We will make such procedures more explicit in the next section.

8.2 The basic design cycle

We will start the formalization of design processes at the level of their building block: the basic design cycle.

A **design cycle** C is a 5-tuple (d, S, K, I, V), where:

d is the **decision** that defines the current design cycle. It indicates the direction which the current step should take. In general it reflects the strategy which the designer adopts to reach the design goal, or some knowledge that influences the current design phase;

S is the **specification** or definition. It is a (formal) statement about the relevant properties of a desired system, such as functionality, behaviour, performance, structure, geometry, etc.;

K is the **knowledge attribute**. It is the knowledge required to derive the implementation from the specification.

I is the **implementation** associated with the current design cycle C.

V is the **verification**. It usually takes the form of some formal proof. For instance, in CCS one were to prove that I and S are equivalent.

The pair (K, I) will also be referred to as the **detailing**. The triple (S, K, I) will also be referred to as the **SKI model** of the design cycle C [Koo84]. [ROL90] analyses the design decisions associated with a program development.

8.3 A design process model

In this section we will develop a simplified view of the design process. In the next section we will refine this view to account for additional real-world situations.

Design involves three domains: (i) the *mind*, (ii) the *formal domain*, and (iii) the *physical domain* (see *Figure 8.1*). The creation of a system starts in someone's mind; there is always a person (or group of persons) who has an idea about a desired system. One might say that the first model of a system is an idea in someone's *mind*. Then comes the problem of putting this idea into a written statement about the system requirements, thereby crossing the boundary between the *mind* and the *formal domain*. The design process involves the evolution of models of the required system from an initial specification in the formal domain to a final implementation in that domain followed by a mapping onto a physical object in the *physical domain*.

In the formal domain we are dealing with a collection of design cycles. The simplest view on the design process is obtained when a system model M is both the implementation of a design cycle and the specification for the next design cycle. In this view a design process yields a series of system models $M_i (0 \leq i \leq n)$ of increasing detail starting with the initial model M_0. The final model M_n should then contain sufficient information such that a physical realization can be derived from it.

We define a system model M as a pair (m, c) where:

(i) m is the *formal system model*

stated in terms of mathematics, programming languages, hardware description languages, lay-out description languages for IC's, etc. Such design languages are used to allow designers to communicate about their designs in an unambiguous way, and also to enable computer support for symbolic reasoning, simulation, analysis, etc.

(ii) c represents the *constraints*

These are informal statements in terms of a natural language, or in terms of pictorial information such as a sketch. This type of information evolves during

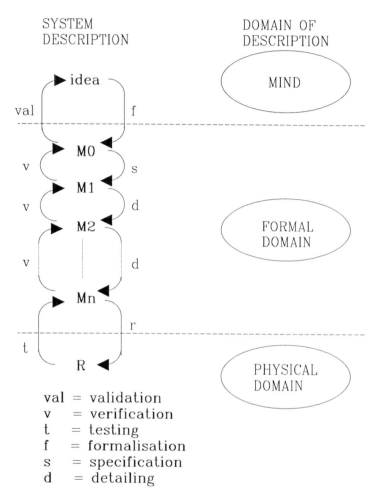

Figure 8.1: The three domains of a design process and the types of detailing and verification steps in it (validation and testing may also be done against an intermediate model $M_n (o \leq i \leq n)$.

the design process. For instance, in the course of the design process the statement
"the system should have a maximum dissipation of 1 Watt", may be transformed
into an electronic structure which minimizes electrical currents. Such informal
information constrains the shaping of the formal information.

The design process is concerned with increasing the level of detail of the formal
model in a number of design cycles, while constraints are gradually turned into
statements of the formal model, yielding the series of models M_i mentioned
earlier. We will use m to denote the final model m_n and c to denote the initial
set of constraints c_0. Usually, the desired system initially is described in a purely
informal way. In that case the constraints reflect the user requirements and
$M_0 = (\varnothing, c)$. At the end of a design process we should have a fully formalized
system satisfying these requirements. Since the constraints are fully met, we can
put $c_n = \varnothing$ and, hence, $M_n = (m, \varnothing)$.

Within our three domain model, we distinguish the following types of design
cycles:

the **formalization step** $(d_f, idea, K_f, (\varnothing, c), V_f)$

This is the design cycle which links the *mind* with the *formal domain*, and involves requirements analysis; d_f is the design decision associated with this step;
K_f represents knowledge about requirement analysis; $(\varnothing, c)$ is the initial model
M_0. V_f represents an informal activity called **validation**, which involves establishing the correspondence between M_0 and the idea in someone's mind that led
to it. Validation can be considered as a special case of verification across the
mind - formal domain boundary.

Once inside the formal domain, a process of refinement and decomposition takes
place. One of the first steps in the process is to obtain a specification of the
desired system in some formal notation (M_0 may be in natural language form).

the **specification step** $(d_s, (\varnothing, c), K_s, (S, c_s), V_s)$

where S is the specification and c_s are the new constraints derived from c (possibly $c = c_s$); d_s is the decision associated with this step (indicating, for example
the kind of specification technique to be used) and K_s represents the design
knowledge about the used specification technique, the knowledge regarding the
required system architecture, etc. Since S cannot be verified formally, V_s reflects
an informal reasoning to show that S satisfies some requirements in c and does
not invalidate the remaining constraints in c. par Subsequent steps are of the
following type:

the i-th **detailing step** $(d_i, M_{i-1}, K_i, M_i, V_i)$.

Notice that the following relations hold:

$$M_0 = (m_0, c_0) = (\varnothing, c), d_0 = d_f, K_0 = K_f, V_0 = V_f;$$
$$M_1 = (m_1, c_1) = (S, c_s), d_1 = d_s, K_1 = K_s, V_1 = V_s;$$
$$M_n = (m_n, c_n) = (m, \varnothing), d_n = d_r, K_n = K_r, V_n = V_r.$$

Formal verification comes into reach whenever two successive models m_i and m_{i+1} are within the same formal domain. For example, if the models are written in terms of CCS, then verification requires m_i and m_{i+1} to be observation equivalent (or some other type of equivalence).

The final design cycle will be referred to as

the **realization step**$(d_r,(m_n, \varnothing),K_r, physical\ object , V_r)$,

where K_r is the knowledge required to perform this step and *physical object* is the physical entity realizing the desired system. In case m_n is a computer program (obtained via a formalization step, a specification step, and one or several detailing steps), then the realization step involves the down loading of the program into a computer and entering the *run* command to start program execution. In case of an integrated circuit, the realization step involves the projection of the masks onto the wafer and the associated chemical processing steps. In general, in case of a hardware design, the realization step involves a process of manufacturing.

Product engineering and process development are both design processes in our terminology and can be described using the model presented here.

V_r reflects a special kind of verification between the *formal domain* and the *physical domain* which is known as **testing**. In case of an integrated circuit, there are two types of testing: *structural testing* (to detect defective gates on the silicon substrate), and *functional testing* (to verify the function of the actual circuit against a simulation model of it). If no verification is done during the design process (either explicitly after each design cycle, or implicitly when correctness-preserving design transformations are carried out), then all the verification burden rests on this testing phase and makes the latter more complicated. This is one of the reasons why testing is such a large activity in industrial development.

The programming example

Let us return to our programming example, and see how this looks in terms of our design process model. *Italics* refer to the text by Gries, whereas expressions between brackets, like <*comment*>, are comments by the author.

The *formalization step* $(d_f, idea, K_f, (\varnothing, c), V_f)$, where

$\mathbf{d}_f = <$ the designer decides that an alternative construct will be used to derive a certain required program $>$;

$\mathbf{K}_f = (14.7)$ and *theorem 10.5* $<$ knowledge about the alternative construct $>$;

$\mathbf{c} =$ "*Write a program that, given fixed integers x and y, sets z to the maximum of x and y*".

Validation requires the inspection of the statement c and convincing oneself that it faithfully reflects the intended functionality. Next comes:

The *specification step* $(d_s, (\varnothing, c), K_s, (S, c_s), V_s)$

with $(\varnothing, c) = M_0$ and $(S, c_s) = M_1$;

d_s = < write the specification in a pseudo programming language >;

K_s = *(14.7)* and *theorem 10.5* < knowledge about the alternative construct >;

S= { T} S { R: z = max (x,y) };

c_s = c;

V_s = $\varnothing$; no verification is done at this point. We proceed to

a *detailing step* $(d_2, M_1, K_2, M_2, V_2)$, where

d_2 = *"Before the program can be developed, R must be refined by replacing* max *by its definition."*

K_2 = *"Variable z contains the maximum of x and y if it satisfies*
$$R : z \geq x \wedge z \geq y \wedge (z = x \vee z = y)".$$

m_2 = $\{T\}S\{R : z \geq x \wedge z \geq y \wedge (z = x \vee z = y)\}$;

c_2 = c.

V_z = < Verification of the post condition is straightforward by inspection >.

We proceed to the next *detailing step* $(d_3, M_2, K_3, M_3, V_3)$, where

d_3 = < refine the program body >;

K_3 = (14.7) and *theorem 16.5*.

m_3 = $\{T\}$ifx $\geq$ y $\rightarrow$ z := xfi$\{R : z \geq x \wedge z \geq y \wedge (z = x \vee z = y)\}$;

c_3 = c.

V_3 = *"This program performs the desired task provided it doesn't abort. (...)* But Q, which is T, does not imply x $\geq$ y."

Hence, verification yields that m_3 is not yet satisfactory. And we need a *detailing step* $(d_4, M_3, (K_4, M_4), V_4)$, where

d_4 = *"... at least one more guarded command is needed."*

K_4 = *(14.7)* and *theorem 16.5*;

$m_4 = \{T\}$ (if $x \geq y \rightarrow z := x \, [\!] \, y \geq x \rightarrow z := y$ fi) $\{R\}$

$c_4 = \varnothing$.

$V_4 =$ *"now, at least one guard is always true, so that this is the required program"*.

8.4 Refinement of the design process model

As was mentioned earlier, the above model is a simplification. To account for real world situations, we will use the following refinements of the model:

(i) Design trees

During a design cycle, a system may be decomposed into subsystems. These, in turn, are subject to design cycles, during which decomposition can occur. Hence, a design cycle in general yields a set of implementations, rather that a single implementation. Each element in the set is then subject to a design cycle, which may lead to a decomposition and corresponding design cycles. In summary, a system decomposition yields a *design tree*, where each node represents a subsystem and the arcs lead to lower level subsystems obtained after decomposition.

(ii) The Ideal Design Process versus the Actual Design Process.

A designer can only work on one design cycle at a time. Moreover, the designer does not develop system models in a straightforward way. Rather, these models are obtained during a process of iterations, each iteration resulting in the updating of the relevant knowledge attribute. Verification either leads to the acceptance of the model (and entering the next design cycle) or the abandoning of it (leading to a new iteration of the current cycle or a previous design cycle); see also *Section 1.1*. Hence, the designer iterates between different levels of design since low level design decisions may have an impact on higher design levels. This may lead to either small scale or large scale iterations.

By the **Actual Design Process (ADP)** we will mean the sequence of design cycles as they are actually carried out by the designer. In case there is a design team, then the ADP is the set of ADPs as carried out by the design team, together with the design tree of the development project (which may be subject to one or several design cycles as well). The ADP may not be the most effective way to document the design process during the design process, several alternatives may have been developed of which only one is selected. The reason why certain alternatives were not acceptable should be reflected in the updating of the design knowledge. Therefore, we will define the **Ideal Design Process (IDP)** as the way the design process is finally documented or perceived.

Experienced designers may carry out ADPs which are very close to the IDP, whereas many more design iterations are required by inexperienced designers. Hence, their ADP may be very different from the IDP.

(iii) Validation and testing

In our simple model, validation takes place in the first design cycle. In principle, however, this activity can be carried out after each design cycle if so desired. This is particularly useful in order to check whether the design corresponds with the idea of the designer or user about the required functionality. Simulation usually is a good way to perform validation, whereas formal verification provides the mathematical rigor required to maintain correctness throughout the design process.

In the testing phase, the physical object is verified against M_n. However, one may also perform testing against more abstract models of the required system.

(iv) Multi-level description of design processes

The mechanisms which were explained in *Section 1.4* can also be applied to design cycles. This gives us the possibility to have design process descriptions at different levels of detail.

For instance, for a design cycle (d, S, K, I, V) we can refine the detailing (K, I) into a sequence of statements $s_1 \ldots s_p$. Each statement $s_j (1 \leq j \leq p)$ is a pair (k_j, M_j), where M_j is some intermediate refinement of the specification S such that finally: $M_p = I$. The attribute k_j is the knowledge attribute associated with the pair of intermediate refinements (M_{j-1}, M_j), with $M_0 = S$. We call the sequence $k_1 k_2 \ldots k_p$ the **line of reasoning** of the design cycle.

We can generalize this into a **step refinement rule**: within a design cycle, each part of a design cycle $(d, S, K, I \text{ or} V)$ can be refined by means of another design cycle. Likewise, we can define a **step decomposition rule**, whereby a whole design cycle is decomposed into a sequence of design cycles (see *Figure 8.2*).

The knowledge attribute $k1$ of a design cycle can itself be obtained through a design cycle where the knowledge specification $k2$ was synthesized into $k1$ using the meta-knowledge $k3$, which in turn was synthesized form the knowledge specification $k4$ using the meta-knowledge $k5$, etc. (see *Figure 8.3*).

The complementary activity of step refinement is **step simplification**. Likewise. the complementary activity of step decomposition is step composition (this can be illustrated by taking the reverse directions in *Figure 8.4*.

8.5   Meta programs

With the remarks from the previous section in mind (in particular those on design trees), we describe a design process as a series of design cycles: $DP = C_1 \, C_2 \ldots C_n$. Using the definition of C, we may write DP as:

$$DP = (d_1, M_0, K_1, M_1, V_1)(d_2, M_1, K_2, M_2, V_2) \ldots (d_n, M_{n-1}, K_n, M_n, V_n)$$

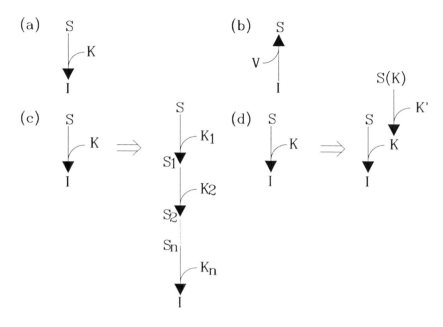

Figure 8.2: (a) Symbol for the SKI model of a design cycle; (b) similar for the verification step; (c) step decomposition; (d) step refinement.

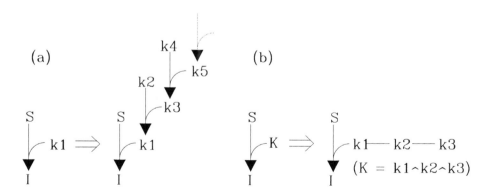

Figure 8.3: Illustration of knowledge refinement and knowledge decomposition: in that latter case we have $K = k1 \wedge k2$.

We can use alternative descriptions, each giving an equivalent view, but with a different focus. For example, if we wish to emphasize the sequences of decisions, models, knowledge attributes and verifications, then we could rewrite DP as:

$$DP_a = (d_1 \ldots d_n)(M_0 M_1 \ldots M_n)(K_1 \ldots K_n)(V_1 \ldots V_n)$$

An alternative would be

$$DP_b = (d_1 \ldots d_n)(M_0 M_1 \ldots M_n)((K_1 V_1) \ldots (K_n V_n)).$$

in which case we emphasize the fact that the knowledge attribute and verification associated with a design cycle belong together; this way of describing a design process is appropriate in constructive approaches.

Since a design decision is concerned with the direction or control of the design process, we may call $(d_1 \ldots d_n)$ the **meta-line of reasoning**, reflecting the decision process associated with the design process. $(M_0 M_1 \ldots M_n)$ is the **design hierarchy** of the design process; (although in our model of DP we use a sequence of models, in reality this usually is a design tree as noted earlier); $(K_1 \ldots K_n)$ is the **line of reasoning** of the design process; $(V_1 \ldots V_n)$ is the **correctness proof** of the design process.

A **meta program** reflects the structure of and the reasoning associated with a design process. It is based on the idea of replacing the models M_i in DP by pointers to the actual models. We define a pointer function p from the integers $\{0 \ldots n\}$ to the models $\{M_0, \ldots M_n\}$, such that $p(i) = M_i$. Then the meta program MP associated with DP is

$$MP = DP[p_i / M_i]$$

where $DP[p_i / M_i]$ means: replace M_i by p_i in the expression for DP.

We introduce a small *Design Process Description Language (DPDL)*. The concatenation of one or more items is indicated using set braces {}. Bold letters denote keywords. Square brackets [ ] indicate optional expressions.

$$< \text{design process} > \quad \Rightarrow \quad \{< \text{step} >\}$$

$$< \text{step} > \qquad\qquad \Rightarrow \quad \textbf{step} < \text{step-id}>$$
$$[\textbf{decision} < \text{decision} >]$$
$$[\textbf{language} < \text{language} >]$$
$$[\textbf{specification} < \text{specification} >]$$
$$\textbf{detailing} < \text{detailing} >$$
$$\textbf{verification} < \text{verification} >$$
$$\textbf{endstep} < \text{step-id} >$$
$$\Rightarrow \quad \textbf{step} < \text{step-id} >$$
$$\{< \text{step}>\}$$
$$\textbf{endstep} < \text{step-id}>$$

$< \text{step-id} >$ is a step identifier.

The step identifier can be structured in such a way that design tree representations are easy to realize.

$< \text{decision} >$ is a text.
$< \text{language} >$ defines the language to be used.

The language or specification parts may be omitted. In that case their default values are given by their last occurrences. The language part may define the language (or languages) to be used for the specification, for intermediate models of the design cycle (obtained via step refinement), and for the implementation of the current design cycle.

An indication about the used language can be particularly interesting in the case of computer support for such languages. For example, using a tool that could read and manipulate the DPDL description, one could select a particular design cycle and automatically enter a language-sensitive editor for the particular type of description.

$$< \text{specification} > \quad \Rightarrow \quad <\underline{\text{specification}} \text{ pointer} >$$
$$\Rightarrow \quad <\underline{\text{specification}} \text{ step} >$$

Underlined words serve as comments; from the point of view of the syntax they have no meaning. Pointers refer to the actual specification or model description. A design cycle may be needed to arrive at the required specification.

$$< \text{detailing} > \quad \Rightarrow \quad [(<\text{lnr}>) < \text{knowledge\_attribute} >]$$
$$<\underline{\text{implementation}} \text{ pointer} >$$
$$\Rightarrow \quad < \underline{\text{detailing}} \text{ step} >$$

$< \text{lnr} >$ is a reference number

A detailing is either a knowledge attribute (preceded by a reference number), followed by a pointer to an implementation. The detailing may also be decomposed into several design cycles.

<knowledge_attribute>   is a (reference to a) description of the knowl-
                         edge attribute.
                         When omitted we say that the result was pos-
                         tulated.

<verification> $\Rightarrow$ (< lnr >) **verify** < verification_condition >
                  $\Rightarrow$ (< lnr >) < verification_statement >
                  $\Rightarrow$ < verification step >

<verification_condition>   a logical expression which should evaluate to
                            true

A verification can be put into some formal notation, or is an informal explanation; alternatively, a design cycle may be needed in order to arrive at the
required verification (for example, certain intermediate results are needed before
the actual verification can be done.

; Comments may always be inserted.
; Each comment line is preceded by a semicolon.

From the syntax description we see that steps may be nested. Nesting can be the
result from refinements or decompositions. For instance, refinement of a design
cycle may yield the following description in DPDL:

step 1
specification
          specification **step** 1.1
          specification < pointer >
          detailing < knowledge_attribute >< pointer >
          specification **endstep** 1.1
detailing
          knowledge **step** 1.2
          specification < pointer >
          detailing(< lnr >) < knowledge_attribute >< pointer >
          knowledge **endstep** 1.2
verification < verification >
endstep 1

Alternatively, we can apply step decomposition:

step 1
          step 1.1
          specification < specification >
          detailing < detailing >
          verification < verification >
          endstep 1.1
          step 1.2
          specification < specification >
          detailing < detailing >

      verification $<$ verification $>$
      endstep 1.2
endstep 1

We will apply these mechanisms in a design method for communicating systems. The next section introduces this method as well as the methodological principles from which the method is derived.

8.6   Design method for communicating systems

A design cycle is based on the derivation of an implementation I from a specification S using certain design knowledge K (the *SKI*-model). Using principles such as step decomposition, step refinement, step composition and step simplification, we can build a hierarchy of descriptions of a design process. Hence, we can apply the *SKI* model to very small design cycles but also to very large ones (a design cycle is considered "small" if its implementation is only slightly more detailed than its specification). We may even view a whole design process as one (very large) design cycle. We postulate the following set of methodological principles:

(a) use a hierarchy of models.

We can use the hierarchy principle to reason about the design process at different levels of detail. Our design process model, the mechanisms such as step refinement. step decomposition, etc, and the language DPDL, automatically yield a hierarchy.

(b) model a system from outside-inwards.

Since we are interested in communicating systems (which are characterized by their interaction with their environment), it seems natural to take a model of a system's environment as the starting point for the design process, while refining the system's inner structure in subsequent steps.

(c) use a mathematical discipline.

Such a discipline enables one to reason about a system's behaviour in a very precise manner; properties can be stated in an unambiguous way and verification can be done formally (and, to some extent, mechanically).

(d) behaviour and structure.

Select a formalism which enables one to describe the *behaviour* as well as the *structure* of a system. There are two major behavioural characteristics: (i) operations on data, and (ii) communication between parts.

Principles (a) and (b) together lead us to the following approach. We will apply the outside-inward principle to the design process viewed as a single design cycle. First, by starting with a model of a system's environment we can obtain a model of the system by obtaining its mirror image (see *Chapter 5* on synthesis by mirroring). This provides a specification of the system which we call the **COMMUNICATION MODEL**. A model of the system's internal structure provides the required knowledge attribute for the design cycle. We will refer to it as the **APPLICATION MODEL**. The implementation will be called the **PROCESS/DATA MODEL**; it combines the external communication behaviour of the required system with its inner structure. Hence, we consider the *IDP* of communicating systems as one single design cycle with the above specification, knowledge attribute and implementation. Then, using step decomposition and step refinements we will obtain more detailed versions of this IDP such that we can use it to model the fine grain design cycles to be carried out.

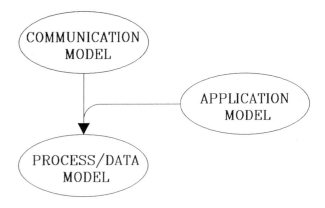

Figure 8.4: The design process as a single design cycle.

With respect to principles (c) and (d), we will select CCS as our mathematical discipline, not only for description purposes but also for reasoning purposes (specifically, verification). The choice was based on the observation that decomposition in a design leads to parts that communicate, while communication lies at the heart of CCS (see also *Section 1.5*). However, operations on data are less efficiently covered by CCS and for this one might decide on another formalism. Structure is reflected in the way subsystems are interconnected, something that is used while applying parallel composition and restriction.

For those, familiar with the Hatley/Pirbhai method [HP87], it may be worthwhile to note the following correspondence with the method presented here. Let us consider each of the methodological principles:

(a) use a hierarchy of models.

Both here, as well as in the book of Hatley and Pirbhai, hierarchy is used to structure design processes. Hierarchies exist in terms of levels of *control flow diagrams*, as well as in terms of *data flow diagrams*. An *architecture flow diagram* shows the relationship (in terms of flows of data objects) between (physical) architecture modules. In our case we will describe such relationships using step decomposition and refinement, and the corresponding communication between CCS agents at different levels.

(b) model a system from outside-inward.

Hatley and Pirbhai's method starts with defining *context flow diagrams*, showing the relationship of a system with its environment. Hence, the above principle is followed. Our *communication model* corresponds with their *context diagram*; refinement and decomposition of the communication model yield a hierarchy of communication models. In [HP87] a hierarchy of *control flow diagrams* is presented, each of which in fact shows the way a "bubble" is linked with its environment.

Pirbhai does not make explicit use of *knowledge attributes*, like we do in this book. In our method, the *application model* expresses this. The explicit structuring of design knowledge is an important and powerful tool to structure and support design processes. Introduction of data will be done here in the *process/data model*, although it can be done also in the *communication model* and the *application model*. In our case we will use *abstract data types* to include data descriptions. Hatley and Pirbhai define data using *requirements dictionaries*, whereas *data flow diagrams* model their flow.

(c) use a mathematical discipline.

The method of Hatley/Pirbhai does not make use of a mathematical formalism. Hence, formal verification is not possible; his hierarchy of *control flow diagrams* is a syntactic method to support consistency between design levels.

(d) behaviour and structure.

In both methods, behaviour and structure are explicitly described. In our method, the possibility to describe an *IDP* or *ADP* in terms of *DPDL* provides a way to describe the structure of design processes explicitly, including the design decisions that lead to a particular design cycle.

In summary, both methods share methodological elements. The method presented here makes use of a mathematical formalism and explicitly models design processes in terms of design cycles. Hatley and Pirbhai's method has some additional structuring elements that are useful in real applications. The reader may find useful structuring elements in both methods that can be combined in actual designs.

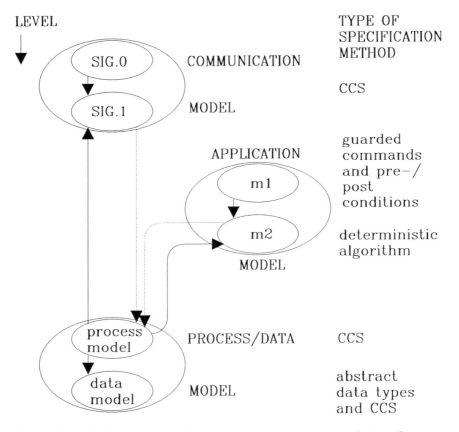

Figure 8.5: Design process of communicating systems and its refinements and decompositions.

Chapter 9

THE COMMUNICATION MODEL

9.1 Introduction

We will consider the design of a communication model of a telecommunication network [Koo85a]. In general, a network consists of *nodes* and *edges.* We say a network is *strongly connected* if there exists a path (sequence of edges) between every pair of nodes in the network. In the case of a telecommunication network, nodes will be referred to as **switching nodes**. A switching node can be a telephone exchange, a node in a packet switching system, a group of telephone exchanges, or switches inside these exchanges.

Each switching node can have **terminal nodes** connected to it. A terminal node is a telephone set, a computer terminal, etc. A particular terminal node can be connected to only one switching node. A terminal node has two modes of behaviour: it either acts as a *source* (when someone makes a telephone call), or it acts as a *destination* (when someone receives a telephone call). When someone wishes to call an another person, then a connection within the network has to be set up between these two persons. Such a connection consists of a sequence of alternating edges and switching nodes between a source terminal node and a destination terminal node.

Each switching node associated with such a connection has to carry out certain actions. The total of actions within a switching node with respect to a particular connection is called a **local connecting process**. The set of cooperating local connecting processes throughout the network required to set up a connection between two subscribers is called the **global connecting process**.

Given the required behaviour of terminal nodes, we can apply the combination algorithm to derive the communication model. Consequently, we first need to describe behaviours of terminal nodes. Such behaviours are constructed from actions, involving the sending and receiving of messages. We will use the symbol

137

s to refer to the *source terminal node*. Likewise, *d* refers to the *destination node*. Finally, *c* refers to the switching node up to which a path has been set up between *s* and *d*. In the sequel we will consider the local connecting process in *c*. We postulate a set *M* of **control messages**:

(1) M = { con, ptr, ntr, ster, dter }, where

con is the **connection request**; this message initializes and instructs the connecting process.

ptr is the **positive terminating response**; this message indicates the successful termination of the connecting process. When received by a node *c* it implies that a path has been set up between *c* and *d*.

ntr is the **negative terminating response**; indicates the failure of the connecting process. When received by a node *c* it implies that no path could be set up between *c* and *d*.

ster is the **source termination request**; this message is sent by the source node *s* to request termination of the connection.

dter is the **destination termination request**; this message is sent by the destination node *d* to request termination of the connection.

We will use integer numbers between brackets (such as **(1)**) to refer to specifications or implementations. We will use numbers preceded by a **k** to refer to knowledge attributes (e.g. **(k1)**).

Using the above messages, we can postulate the behaviours of the source node *s* and the destination node *d*. These behaviours will be denoted as t_s and t_d respectively. Let *G* denote the network of switching nodes. Let *s.in* and *s.out* denote the input and output ports of the source *s* towards *G*. Likewise, *d.in* and *d.out* are the input and output ports respectively of the destination *d* towards *G*. We postulate the following behaviours for t_s and t_d:

(2)t_s ::S0	= s.out!con	: S1	**(3)**t_d :: D3	= d.in?ster	: D0
S1	= s.out!ster	: S0		+ d.out!ptr	: D4
	+ s.in?ptr	: S2		+ d.out!ntr	: D0
	+ s.in?ntr	: S5	D4	= d.in?ster	: D6
S2	= s.out!ster	: S0		+ d.out!dter	: D0
	+ s.in?dter	: S5	D6	= d.out!dter	: D0
S5	= s.out!ster	: S0			

9.2 Applying the combination algorithm

Step 1 of the algorithm (write behaviours in normal form) has already been carried out. Step 2 requires the following mirroring operation:

(k1) (*mirroring*)
M$_1$ :: s.out→ G.in1 M$_2$ :: d.out→ G.in2
s.in→ G.out1   d.in→ G.out2

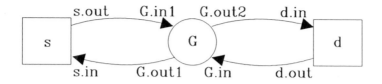

Figure 9.1: Communication ports between the network G and the source and destination.

The following renaming of behaviour identifiers will also be carried out:

S0	$\rightarrow$	G0	and	D0	$\rightarrow$	G0
S1	$\rightarrow$	G1		D3	$\rightarrow$	G3
S2	$\rightarrow$	G2		D4	$\rightarrow$	G4
S5	$\rightarrow$	G5		D6	$\rightarrow$	G6

Carrying out the mirror operations $G_d = M_1(t_s)$ and $G_s = M_2(t_d)$ yields:

(4) G_d :: G0 = G.in1? con   : G1
            G1 = G.in1? ster   : G0
                + G.out1! ptr   : G2
                + G.out1! ntr   : G5
            G2 = G.in1? ster   : G0
                + G.out1 dter   : G5
            G5 = G.in1? ster   : G0

(5) G_s :: G0 = G.out2! con   : G3
            G3 = G.out2! ster   : G0
                + G.in2? ptr   : G4
                + G.in2? ntr   : G0
            G4 = G.out2! ster   : G6
                + G.in2? dter   : G0
            G6 = G.in2? dter   : G0

Step 2 of the algorithm also requires a choice of the initial state.

(k2) *(knowledge of initial state)* ::

The initial state of the combined system consists of the initial states of its sub-systems. Hence, we will use the pair $(G0, G0)$ as the initial state, where the left hand $G0$ is the initial state of G_d; the right hand $G0$ is the initial state of G_s.

In step 4 of the algorithm, causal relations need to be defined. To derive these relations we will use the following knowledge attribute:

(k3) *(knowledge of causal relations)* ::

a message m is part of a causal relation if it occurs both within an input action
and within an output action of the same system; the direction of causality (indicated by "$=\rightarrow$") is from input action to corresponding output action; see also
the **Synth** rules of *Section 5.3.2.*

In the interface synthesis as treated in *Section 5.3*, it was assumed that messages,
when arrived at the interface system, could immediately result in the sending or
receiving of messages, causally related to these. In fact, we have implicitly assumed that the delay within the interface system is negligible compared to the
communication with the external world. In the case of a telecommunication network, however, such an assumption may not be valid. Hence, we have to take
into account a possible delay within the network. We can do this by introducing
the **star version of an action**. For instance, the star version of a message m,
denoted as m^*, would indicate that message m has been transferred through the
network; in that case it is available to effectuate the causal relation involving
the sending or receiving of the causally related message at the required location
within the network. For instance, if we make a telephone call and send a connection request (con) to the network, then the message con^* is required at the
switching node connected to the destination terminal node before the ringing
signal (also modeled as a con message) can be sent to that terminal node. The
expression $con \rightarrow con^*$ denotes the availability of the con message at the required
switching node. Using this, we can formulate the following knowledge attribute:

(k4) *(knowledge about internal delay in the network)* ::

Since we assume internal delay in the network, the causality relations have to
account for this by a condition reflecting the internal transfer of messages using
the above type of expression.

The resulting set of causal relations is:

(k5) *(definition of causal relations* CAUS $(G_s, G_d))$ ::

(a_1)	G.in1?con	and	(con	$\rightarrow$	con^*)	$=\rightarrow$	G.out2!con
(a_2)	G.in1?ster	and	(ster	$\rightarrow$	$ster^*$)	$=\rightarrow$	G.out2!ster
(b_1)	G.in2?ptr	and	(ptr	$\rightarrow$	ptr^*)	$=\rightarrow$	G.out1!ptr
(b_2)	G.in2?ntr	and	(ntr	$\rightarrow$	ntr^*)	$=\rightarrow$	G.out1!ntr
(b_3)	G.in2?dter	and	(dter	$\rightarrow$	$dter^*$)	$=\rightarrow$	G.out1!dter

(k6) *(knowledge of combination algorithm)* ::

Using the equations G_d and G_s, the relations $CAUS(G_s, G_d)$ as well as the
definition of $=\rightarrow$, we can apply steps 5-10 of the combination algorithm. This
yields a set of 49 equations:

(6) G_{sd} :: see *Appendix B.*

When observing these equations, one notices that certain derivations carry the
label *axiom*. The explanation is as follows. During the application of the algorithm we encounter situations where certain messages are still pending, although

these messages counteract each other. For instance, suppose the network has returned to its initial state, while the messages *con* and *ster* are both pending. In that case, it seems useless to first transmit *con* through the network, thereby setting up a connection, and then removing that connection again due to the *ster* message. It is much more effective to have these messages cancel each other as soon as it is detected that they are both pending. This requires an additional design decision by the network designer.

Therefore, throughout the application of steps 5-10 of the combination algorithm, the designer may decide that straightforward application of these steps should not occur; instead, the designer decides that certain steps should not occur, or certain messages should be replaced by other messages. For example, a *ptr* message, directly followed by a *dter* message could be replaced by an *ntr* message, since it makes no sense to first acknowledge the setting up of the connection, immediately followed by the braking up of this same connection.

These considerations yield the following knowledge attributes **k7** and **k8**:

(k7) *(knowledge by the designer about unuseful message sequences)* ::

When the message queue is not empty while the system is in its initial state, then the following message pairs in the queue are replaced:

con,ster	$\rightarrow$	(*con* and *ster* cancel each other)
con*,ster*	$\rightarrow$	(same for their *star versions*)
ster*,ntr	$\rightarrow$	(*ster** and *ntr* cancel each other)
ster,ntr*	$\rightarrow$	(*ster* and *ntr** cancel each other)
ptr,dter	$\rightarrow$ ntr	(*ptr* and *dter* interpreted as *ntr*)
ptr*,dter*	$\rightarrow$ ntr*	(similar for their *star versions*)

The above replacements yield a set of design decisions which we will refer to as *axioms*, since they are directly generated by the designer and not derived within the formal derivation. These axioms are summarized in the following knowledge attribute (numbers refer to equations in *Appendix B*):

(k8) *(axioms)*

AX-1 ::

G0G0 (con,ster)	= G0G0	(3)
G0G0 (con*,ster*)	= G0G0	(9)
G0G0 (ster*,ntr)	= G0G0	(19)
G0G0 (ster,ntr*)	= G0G0	(20)
G1G0 (ptr,dter)	= G1G0 (ntr)	(21)
G1G0 (ptr*,dter*)	= G1G0 (ntr*)	(24)
G0G0 (ster,ptr,dter)	= G0G0 (ster,ntr)	(25)
G0G0 (ster*,ptr,dter)	= G0G0 (ster*,ntr)	(26)
G0G0 (ster,ptr*,dter*)	= G0G0 (ster,ntr*)	(29)
G0G0 (ster*,ptr*,dter*)	= G0G0 (ster*,ntr*)	(30)
G0G0 (ptr,dter)	= G0G0	(34)
G0G0 (ptr*,dter*)	= G0G0	(37)

In addition, the following design decisions were taken:

AX-2 ::

G0G0 (ster*,ntr*)	= G0G0	(31)
G0G0 (ntr)	= G0G0	(35)
G0G0 (ster*)	= G0G0	(40)

We are now ready to execute the last step of the algorithm:

(k9) *(knowledge of combination algorithm and CCS)* ::

The equations for G_{sd} can be reduced under the application of the CCS laws. This yields (for the detailed steps see *Appendix C*):

(7) G_{sd} (reduced) ::

G0 =	G.in1?con	: G1		G6 =	G.in1?ster	: G0
G1 =	G.in1?ster	: G2			+ G.out1!ntr	: G11
	+ G.out2!con	: G3		G7 =	G.out2!ster	: G12
G2 =	G.out2!con	: G4			+ G.in2?dter	: G0
	+ τ	: G0		G8 =	G.in1?ster	: G7
G3 =	G.in1?ster	: G4			+ G.in2?dter	: G9
	+ G.in2?ptr	: G5		G9 =	G.in1?ster	: G0
	+ G.in2?ntr	: G6			+ G.out1!dter	: G11
G4 =	G.out2!ster	: G0		G10 =	G.in1?ster	: G0
	+ G.in2?ptr	: G7			+ G.out1!ptr	: G9
	+ G.in2?ntr	: G0			+ τ	: G6
G5 =	G.in1?ster	: G7		G11 =	G.in1?ster	: G0
	+ G.out1!ptr	: G8		G12 =	G.in2?dter	: G0
	+ G.in2?dter	: G10				

During this reduction the following renaming of agent identifiers took place:

(k10) *(renaming of agent identifiers)* ::

G0G0	→ G0	G0G4 (ster*,ptr*)	→ G7
G1G0 (con*)	→ G1	G2G4	→ G8
G0G0 (con*,ster)	→ G2	G2G0 (dter)	→ G9
G1G3	→ G3	G1G0 (ptr*,dter)	→ G10
G0G3 (ster*)	→ G4	G5G0	→ G11
G1G4 (ptr*)	→ G5	G0G6	→ G12
G1G0 (ntr*)	→ G6		

9.3   Introducing blocking behaviour

From the specification of **(3)** t_d we observe that the destination is the origin of the *ntr* message. However, we will refine the previous result by assuming that the network itself may generate an *ntr* message due to the limited processing capacity of the switching nodes and the limited number of connections that can be sustained at any given moment. In telephony terms this is called *blocking*. Blocking may also occur due to failure; a particular link between two nodes may fail because of natural causes (like earthquakes) or man-made causes (e.g. by digging).

Receipt of a connection request by the network either results in its arrival at the destination or leads to an *ntr* due to blocking. Hence, to account for blocking we will introduce the following causal relation:

(k11) *(knowledge about blocking being introduced into causal relations)* ::

(c_1)(G.in1?con $=\rightarrow$ (con $\rightarrow$ con$*$))$\underline{exor}$(G.in1?con $=\rightarrow$ G.out1!ntr)

(k6) *(knowledge of combination algorithm)* ::

Re-applying the combination algorithm with the additional relation c_1 yields:

(8) G_{sd} with expressions (2) and (4) replaced by:

G5G0 (con) =	G.in1?ster	: G0G0 (con,ster)
	$+ \tau$	: G1G0 (con*)
	+ G.out1!ntr	: G5G0 (con)
G5G0 (con) =		: G5G0
G1G0 (con*) =	G.in1?ster	: G0G0 (con*,ster)
	+ G.out2!con	: G1G3
	$+ \tau$	: G1G0 (ntr*)

In the new derivation two additional axioms were used:

(k12) *(new axioms required due to introduction of blocking)* ::

G5G0 (con)	=	G5G0
G1G0 (con*)	$\overset{\tau}{\longrightarrow}$	G1G0 (ntr*)

As in the original case, reduction of equations **(8)** G_{sd} (with blocking) takes place on the basis of CCS laws:

(k9) *(knowledge of combination algorithm and CCS)* ::

This yields (see Appendix D):

(9) Equations **(7)** G_{sd} (reduced) with $G0$ as follows:
G0 = G.in1?con : (G1 $+ \tau$: G6).

9.4   Verification of the result

Verification means to show that the network, when communicating with the source, shows the correct behaviour towards the destination and vice versa. Stated formally:

(10) (a) $C(G_{sd} \mid t_d)\backslash P(G,d) \approx^m t_s$.   (b) $C(t_s \mid G_{sd})\backslash P(s,G) \approx^m t_d$.

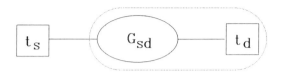

Figure 9.2: Verification means showing mirror observation equivalence between $C(G_{sd}, t_d)$ and t_s; similiar for $C(t_s, G_{sd})$ and t_d.

Calculating expressions **(10)** proceeds in the following steps. Define the following relabellings:

(k13) *(relabellings)*

R_s::			R_d::	
s.out!con	$\to$ c1!		d.in?con	$\to$ c1?
s.out!ster	$\to$ c2!		d.in?ster	$\to$ c2?
s.in?ptr	$\to$ c3?		d.out!ptr	$\to$ c3!
s.in?ntr	$\to$ c4?		d.out!ntr	$\to$ c4?
s.in?dter	$\to$ c5?		d.out!dter	$\to$ c5!
GR_s::			GR_d::	
G.in1?con	$\to$ c1?		G.out2!con	$\to$ c1!
G.in1?ster	$\to$ c2?		G.out2!ster	$\to$ c2!
G.out1!ptr	$\to$ c3!		G.in2?ptr	$\to$ c3?
G.out1!ntr	$\to$ c4!		G.in2?ntr	$\to$ c4?
G.out1!dter	$\to$ c5!		G.in2?dter	$\to$ c5?

Applying these relabellings to t_s, t_d, and G_{sd} yields:

(k14) *(relabelling of source)*   **(k15)** *(relabelling of destination)*

$SRCE = t_s[R_s]$ ::			$DEST = t_d[R_d]$ ::		
S0 =	c1!	: S1	D0 =	c1?	: D1
S1 =	c2!	: S0	D1 =	c2!	: D0
	+ c3?	: S2		+ c3!	: D2
	+ c4?	: S3		+ c4!	: D0
S2 =	c2!	: S0	D2 =	c2?	: D3
	+ c5?	: S3		+ c5!	: D0
S3 =	c2!	: S0	D3 =	c5!	: D0

(k16) *(relabelling of network at source side)*

$SG_{sd}=$ $G_{sd}[GR_s]$::

G0 =   c1!: G0a
G0a =   G1 + τ : G6
G1 =   c2?: G2 + G.out2!con: G3

$$
\begin{aligned}
\text{G2} &= && \text{G.out2!con: G4} + \tau : \text{G0} \\
\text{G3} &= && c2?: \text{G4} + \text{G.in2?ptr} : \text{G5} + \text{G.in2?ntr} : \text{G6} \\
\text{G4} &= && \text{G.out2!ster} : \text{G0} + \text{G.in2?ptr} : \text{G7} + \text{G.in2?ntr} : \text{G0} \\
\text{G5} &= && c2? : \text{G7} + c3!: \text{G8} + \text{G.in2?dter} : \text{G10} \\
\text{G6} &= && c2?: \text{G0} + c4! \ \text{G11} \\
\text{G7} &= && \text{G.out2!ster} : \text{G12} + \text{G.in2?dter} : \text{G10} \\
\text{G8} &= && c2?: \text{G7} + \text{G.in2? dter} : \text{G9} \\
\text{G9} &= && c2?: \text{G0} + c5!: \text{G11} \\
\text{G10} &= && c2?: \text{G0} + c3!: \text{G9} + \tau : \text{G6} \\
\text{G11} &= && c2?: \text{G0} \\
\text{G12} &= && \text{G.in2? dter: G0}
\end{aligned}
$$

(k17) *(relabelling of network at destination side)*

$$
\begin{aligned}
\text{DG}_{sd} &= && \text{G}_{sd} \ [\text{GR}_d]:: \\
\text{G0} &= && \text{G.in1? con : G0a} \\
\text{G0a} &= && \text{G1} + \tau : \text{G6} \\
\text{G1} &= && \text{G.in1?ster} : \text{G2} + c1! : \text{G3} \\
\text{G2} &= && c1! \ \text{G4} + \tau : \text{G0} \\
\text{G3} &= && \text{G.in1?ster} : \text{G4} + c3? : \text{G5+c4?} : \text{G6} \\
\text{G4} &= && c2! : \text{G0} + c3?: \text{G7} + c4? : \text{G0} \\
\text{G5} &= && \text{G.in1?ster} : \text{G7} + \text{G.out1ptr} : \text{G8} + c5?: \text{G10} \\
\text{G6} &= && \text{G.in1?ster} : \text{G7} + \text{G.out1!ptr} : \text{G8} \\
\text{G7} &= && c2! \ \text{G12} + c5? : \text{G0} \\
\text{G8} &= && \text{G.in1? ster} : \text{G7} + c5? : \text{G9} \\
\text{G9} &= && \text{G.in1?ster} : \text{G0} + \text{G.out1!dter} : \text{G11} \\
\text{G10} &= && \text{G.in1?ster} : \text{G0} + \text{G.out1!ptr} : \text{G9} + \tau : \text{G6} \\
\text{G11} &= && \text{G.in1? ster} : \text{G0} \\
\text{G12} &= && c5!: \text{G0}
\end{aligned}
$$

We can now write equations **(10)** as follows:

(11)   (a) $(\text{SRCE}| \ \text{SG}_{sd})\backslash\{c1,c2,c3,c4,c5\} \quad \approx^m t_d$
       (b) $(\text{DG}_{sd} \ |\text{DEST})\backslash\{c1,c2,c3,c4,c5\} \quad \approx^m t_s$

(k18) *(knowledge of the expansion law)* ::

Applying the expansion theorem yields:

(12) $(\text{SRCE}| \ \text{SG}_{sd})\backslash\{c_1, c_2, c_3, c_4, c_5\}$ ::

$$
\begin{aligned}
\text{S0G0} &= && \tau : \text{S1G0}_a \\
\text{S1G0}_a &= && \text{S1G1} + \tau : \text{S1G6} \\
\text{S1G1} &= && \tau : \text{S0G2} + \text{G.out2!con} : \text{S1G3} \\
\text{S1G6} &= && \tau : \text{S0G0} + \tau : \text{S3G11} \\
\text{S0G2} &= && \text{G.out2!con} : \text{S0G4} + \tau : \text{S0G0} \\
\text{S1G3} &= && \tau : \text{S0G4} + \text{G.in2?ptr} : \text{S1G5} + \text{G.in2? ntr} : \text{S1G6}
\end{aligned}
$$

S3G11 = τ : S0G0
S0G4 =   G.out2!ster : S0G0 + G.in2?ptr : S0G7 + G.in2?ntr : S0G0
S1G5 = τ : S0G7+ G.in2?dter : S1G10 + τ : S2G8
S1G10 = τ : S0G0 + τ : S2G9 + τ : S1G6
S0G7 =   G.out2!ster : S0G12 + G.in2?dter : S0G0
S2G8 = τ : S0G7 + G.in2?dter : S2G9
S2G9 = τ : S0G0 G9D0 = G.in1?ster : G0D0
S0G12 =   G.in2?dter : S0G0

(13) $(DG_{sd} \mid DEST)\backslash\{c_1, c_2, c_3, c_4, c_5\}$ ::
G0D0 =   G.in1?con : G0$_a$d0
G0$_a$D0 = G1D0 + τ : G6D0
G1D0 =   G.in1?ster : G2D0 + τ : G3D1
G6D0 =   G.in1?ster : G0D0 + G.out?ntr : G11D0
G2D0 = τ : G4D1 + τ : G0D0
G3D1 =   G.in1?ster : G4D1 + τ : G5D2 + τ : G6D0
G11D0 =   G.in1?ster : G0D0
G4D1 = τ : G0D0 + τ : G7D2 + τ : G0D0
G5D2 =   G.in1?ster : G7D2 + G.out1!ptr : G8D2 + τ : G10D0
G7D2 = τ : G12D3 + τ : G0D0
G8D2 =   G.in1?ster : G7D2 + τ : G0D0
G10D0 =   G.in1?ster : G0D0 + G.out1!ptr : G9D0+ τ : G6D0
G12D3 = τ : G0D0 + G.out1!dter : G11D0

(k9) *(knowledge of combination algorithm and CCS)* ::

Equations **(12)** can be reduced using the CCS laws.

(k19):: supplemented with the fairness rule **Fair-2**. This yields:

(14) S0G0 = τ : S0G2
    S0G2 = G.out2!con : S0G4
    S0G4 = G.out2!ster :S0G0 + G.in2?ptr : S0G7 + G.in2?ntr : S0G0
    S0G7 = G.out2!ster :S0G12 + G.in2?dter:S0G0
    S0G12 = G.in2?dter : S0G0

(k20) *(knowledge of mirror observation equivalence)* ::

We conclude that equations **(14)** are mirror observation equivalent with t_d.

(k9) *(knowledge of combination algorithm and CCS)* ::

Equations **(13)** can also be reduced using the CCS laws. This yields:

(15) G0D0 = G.in1?con: G10D0
    G6D0 = G.in1?ster : G0D0 + G.out1!ntr : G11D0
    G11D0 = G.in1?ster : G0D0
    G10D0 = G.out1!ptr : G9D0 + τ : G6D0
    G9D0 = G.in?ster : G0D0 + G.out1!dter : G11D0

(**k20**) *(knowledge of mirror observation equivalence)* ::

We conclude that equations (**15**) are mirror observation equivalent with t_s.

9.5   The meta program

Using the symbolism introduced in *Chapter 8* we will give the meta program for each of the major steps:

$$(a)(b)$$
$$\downarrow\!\!\!- k$$
$$(c)$$

denotes that the model, associated with the label (c) is derived from (a) and (b) on the basis of certain knowledge designated by the label (k).

First step: *mirroring.*

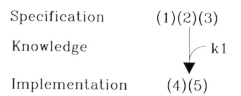

Specification   (1)(2)(3)

Knowledge $\downarrow\!\!\!- k1$

Implementation   (4)(5)

Second step: *applying the combination algorithm:*

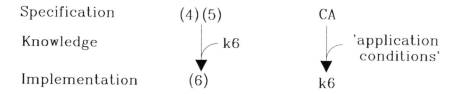

Specification   (4)(5)   CA

Knowledge $\downarrow\!\!\!- k6$ $\downarrow\!\!\!-$ 'application
                                                               conditions'

Implementation   (6)   k6

CA refers to knowledge about the Combination Algorithm; "application conditions" refer to adapting (the initial settings of) the algorithm to the specific application at hand. Again, we can refine this meta-knowledge.

Let CA/IC refer to generic knowledge about the initial condition (initial state) from which the algorithm has to start. Let $CA/CAUS$ denote generic knowledge about causality relations. Finally, let CA/AX denote such knowledge concerning

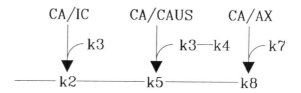

the use of axioms during the expansion phase of the algorithm. Then *application conditions* can be synthesized in two steps. First, we apply decomposition followed by refinement:

Third step: *reducing equations.*

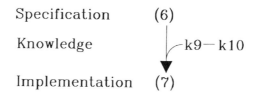

Using step composition we find the complete meta program. *Figure 9.3* shows this meta program as well as a simplified version after using knowledge and step simplification.

Introducing blocking behaviour leaves most of the meta program as it is; changes are made in **k5** (a new causal relation **k11** is added) and **k8** (new axioms **k12** are added) yielding the new results **(8)** and **(9)** (*Figure 9.4*). *Figure 9.5* shows one of the meta programs for verification.

Exercise 9.1 Derive the meta programs for the verification   ∎

9.6 Communication model of switching nodes

Up till now we described the behaviour of the network towards a source and a destination as if it were a single node. However, we still have to derive the behaviour of the individual switching nodes. Hence, the decomposition of the network into switching nodes leads to an additional detailing step where the behaviour of the network is synthesized into the behaviours of its constituent nodes. We will base this detailing step on the following knowledge:

(k21) *(knowledge about the topology of switching networks)* ::

A path in the network, between a source node s and a destination node d, is a sequence of switching nodes. We can refine this knowledge as follows:

(k22) *(knowledge about delay within switching nodes)* ::

We will assume delay will occur between as well as inside nodes. We will model this delay, like in the case of the network, as a FIFO.

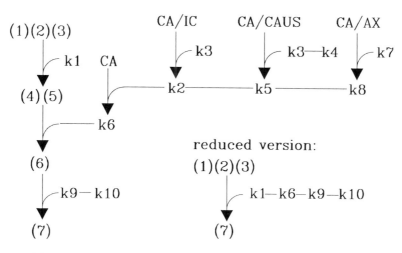

Figure 9.3: Meta program for the derivation of the communication model of the network of switching nodes; reduced version and full program.

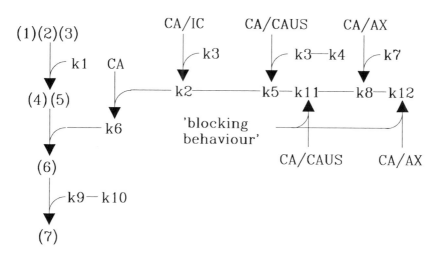

Figure 9.4: Adapting the meta program to account for blocking behaviour of the network.

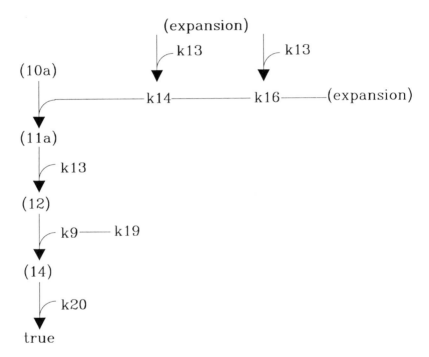

Figure 9.5: Meta program of the first verification; the second verification proceeds in a familiar manner.

In deriving the behaviour of switching nodes, we make use of the following observation:

(k23) *(awareness that there exist certain symmetries)* ::

The behaviour of a switching node can be obtained in a way very similar to the derivation of (7) G_{sd} by observing that the behaviour of a node towards its predecessor on the path is that of the next node relative to its predecessor, whereas it shows source behaviour towards its successor. This is illustrated in *Figure 9.6.*

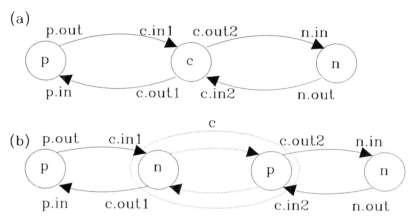

Figure 9.6: (a) the *current node* c (i.e. the node up to which a path has been established from the source node), its predecessor or *previous node* p and its successor or *next node* n; (b) the current node seen as next node of p and previous node of n.

We make the following design decision:

(k24) *(re-phrasing k23)* ::

The behaviour of p is that of t_s; the behaviour of n is its mirror image (i.e. that of G_d).

Let P and N denote the p-type and n-type behaviour of the current node. We next define the following relabellings:

(k25) M_p:: n.out → c.in2   (k26) M_n:: G.in1 → c.in1
            n.in → c.out2   G.out1 → c.out1

Then P and N are defined by:

(16) $P = t_s[M_p]$ (17) $N = G_d[M_n]$

From these equations, the derivation of the switching node behaviour C (for *Current node behaviour*) follows along a similar meta-program as that for G_{sd}. Also blocking behaviour can be likewise included. The resulting behaviour is:

(18)   C::

C0 =	c.in1?con	: C0$_a$		C5 =	c.in1?ster	: C7
C0$_a$ =	τ : C0$_b$ + τ	: C1			+ c.out1!ptr	: C8
C0$_b$ =	c.in1?ster	: C0			+ c.in2?dter	: C10
	+ c.out1!ntr	: C0$_c$		C6 =	c.in1?ster	: C11
C0$_c$ =	c.in1?ster	: C0			+ c.out1!ntr	: C12
C1 =	c.out2!con	: C3		C7 =	c.out2!ster	: C0
	+ c.in1?ster	: C2			+ c.in2?dter	: C11
C2 =	c.out2!con	: C4		C8 =	c.in1?ster	: C7
	+ τ	: C0			+ c.in2?dter	: C9
C3 =	c.in1?ster	: C4		C9 =	c.in1?ster	: C11
	+ c.in2?ptr	: C5			+ c.out1!dter	: C12
	+ c.in2?ntr	: C6		C10 =	c.in1?ster	: C11
C4 =	c.out2!ster	: C0			+ c.out1!ptr	: C9
	+ c.in2?ptr	: C7			+ τ	: C6
	+ c.in2?ntr	: C11		C11 =	c.out2!ster	: C0
C12 =	c.in1?ster	: C11				

Verification proceeds to show that

(19) $(C| \ C) \setminus P_c \approx^m M(C)$

where P_c denotes the port set connecting two switching nodes, both with behaviour C. In fact, (18) is the induction step needed to prove that a sequence of switching nodes, each with behaviour C, behave like a single switching node (with behaviour C).

However, since $C \neq G_{sd}$ we also conclude that the *last* node (i.e. the switching node to which the destination is connected) should have the behaviour G_{sd} instead of C. To verify the correctness of this conclusion it can be shown that

(20) $(C| \ G_{sd}) \setminus P(C,G_{sd}) \approx^m M(G_{sd})$

where $P(C, G_{sd})$ is the port set connecting the sequence of switching nodes (except the last one) with the last switching node.

Chapter 10

THE APPLICATION MODEL

10.1   The formalization step

In *Chapter 9* we developed the communication model as the specification of
our design process aimed at deriving a switching node implementation. The
design process focusses on the control aspect of a telecommunication network
with respect to the establishment of connections between (terminal) modes. In
this chapter we will derive the required knowledge attribute for the above design
process. We will refer to this knowledge attribute as the **application model**,
since it describes in detail our knowledge about the type of application we have in
mind: establishing connections between terminal nodes in a telecommunication
network.

The application model describes the basic structure of the telephony application
rather than its external behaviour, which was modeled in the communication
model. Since our application will be the establishment of connections between
subscribers, we will focus on the call processing function in a communications
network. However, bear in mind that a telephone exchange performs many additional functions such as *call-forwarding, follow-me, cost metering, automatic
re-dial*, and many others.

The first step in deriving the application model will be the formalization step.
The specification of this step will be our general understanding of what it means
to establish connections between terminal nodes in a telecommunication network.
The decision associated with this formalization step will be as follows. Since
call processing is at the heart of switching systems, we choose to derive the
application model as a routing algorithm. In order to derive such a routing
algorithm, we need to have some concepts which enable us to describe the setting
up of connections in networks. Since we need to define suitable concepts to
describe routing in telecommunication networks, we need a detailing step for the
knowledge attribute of the formalization step. Hence:

(1) *first model* ::

S_f = { *general understanding of what it means to establish connections between terminal nodes in a telecommunication network* }.

d_f = { *we derive the application model as a routing algorithm, for which we need suitable concepts; we perform a knowledge step to derive the required knowledge attribute* }.

The specification for the knowledge step can be formulated as follows:

(k1) = { *derive suitable network concepts that will enable us to describe routing in telecommunication networks* }

We decide to organize the detailing step for the specification **(k1)** of the knowledge attribute of the formalization step as follows:

decision for **(k1)**:
We will derive the required concepts in a number of smaller steps. In the first step, we will derive general graph theoretic constructs. In the second step, we will define three node sets, which we will transform into routing predicates. We can then use these predicates as knowledge attribute for the formalization step, leading to a routing algorithm.

The meta program of the formalization will have the following structure (*italics* are descriptions of the relevant DPDL elements):

<u>formalization</u> **step**
decision d_f
specification S_f
detailing
   <u>knowledge</u> **step**
   specification (k1)
   detailing
   { *definition of network topologies* }   ;description of the
   { *definition of "service area"* }   ;knowledge attributes which
   { *definition of three node sets* }   ;have to be derived
   <u>implementation</u> { *routing predicates* };after which we obtain predicates
   endstep
<u>implementation</u> { *routing algorithm* }   ;formalization ends with algorithm
endstep   ;end of formalization step

10.2   Network concepts

We start our formalization by introducing some graph-theoretic concepts, some of which were taken from [Deo74]. Each formal definition will be preceeded by an informal definition of the concept. In the sequel, the contents of this section will be referred to as knowledge attribute **(k2)**.

(k2) ::

A graph consists of nodes and edges. A *directed graph* is a graph in which each edge has a direction:

Definition 10.1
A **directed graph G** is defined by the pair (X, V), where:

-   X is a set of elements called **nodes**, and

-   V is a family of elements from $X \times X$, called **edges** (an element (x, y) from $X \times X$ can appear more than once in this family to indicate that between two nodes more than one edge may exist; in that case we call V a *bag* of edges). The **direction** of an edge (x, y) is from x to y.

In the illustration of *Figure 10.1a*, $X = \{a, b, c\}$ and $V = \{(a, b), (b, c), (c, a)\}$.

We will assume our graphs to be **strongly connected**. Strongly connectedness means that there exists directed paths (i.e. sequences of edges) between any pair of nodes in both directions; this is a prerequisite for being able to make connections between any pair of nodes.

We can think of the nodes X to be *switching nodes*. Since we want to speak about both switching nodes (such as telephone exchanges) and terminal nodes (such as telephone sets), we need to extend our definition to include two types of nodes. In graph theory, the concept of *bipartite graph* partly covers what we want (it defines a graph in terms of two node sets and edges between nodes from different sets, but not between nodes from the same set):

Definition 10.2
A **bipartite (directed) graph BG** is defined by the triple (X, Y, W), where:

-   X is the above set of nodes;

-   Y is another set of nodes, such that $X \cap Y = \emptyset$;

-   W is a family of edges from $(X \times Y) \cup (Y \times X)$
    (these are the edges linking two nodes from different sets).

In the illustration of *Figure 10.1b*, X is as before, $Y = \{d, e, f\}$ and $W = \{(a, e), (e, b), (c, f)\}$.

We have to add the edges between nodes of X to obtain what we want:

Definition 10.3
A **semi-bipartite graph SBG** is defined by the pair (BG, V), where:

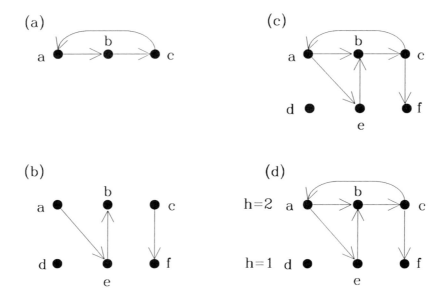

Figure 10.1: illustration of (a) a directed graph; (b) a bipartite graph; (c) a semi-bipartite graph; (d) a heterarchical network.

- BG is the bipartite graph (X, Y, W), and

- V is the set of edges defined in *definition 10.1*.

SBG is illustrated in *Figure 10.1c*.

Notice that from this definition it follows that the class of bipartite graphs is a subclass of the class of semi-bipartite graphs and is defined by $(BG, \varnothing) = BG$. It follows also that the class of directed graphs is a subclass of semi-bipartite graphs and is defined by $((X, \varnothing, \varnothing), V) = (X, V)$.

There is an implicit hierarchy between the switching nodes in telecommunication networks (e.g. those giving access to countries, to areas, to districts, to cities, etc.). Therefore we need to extend our definition with the notion of a *hierarchy*. Since we want to be able to describe networks that contain hierarchies as well as nodes which bear no hierarchical relationships, we will use the term *heterarchical networks* to describe such a mixture:

Definition 10.4
A **heterarchical network** is defined as a pair (SBG, h) where:

- SBG is the semi-bipartite graph (BG, V) with $BG = (X, Y, W)$;

- h is the **hierarchy function**: $X \cup Y \rightarrow \mathbb{N}$, which maps nodes from $X \cup Y$ onto natural numbers. The image $h(x)$ of a node x is called the **order** of x.

(SBG, h) is illustrated in *Figure 10.1d.*

A telecommunication network carries two types of information: *(i)* user information such as speech, and *(ii)* control information such as dialling information. We will restrict ourselves to the latter type of information, since it is control information which drives the process of establishing connections between terminal nodes (in computer terminology, we will focus on the control path and not on the data path). Dealing with user information involves, amongst other things, the design of transmission lines, coding or modulation schemes and detection and receiver equipment, and is not the subject of this book.

Control information is processed and transferred by what we will refer to as the *telecommunication control network*:

Definition 10.5
A **telecommunication control network** G is a heterarchical network (SBG, h), where:

SBG = ((X,Y,W),V);
X: the set of switching nodes;
Y: the set of terminal nodes;
W: the family of edges between switching nodes X and terminal nodes Y:
$$W \subset (X \times Y) \cup (Y \times X);$$
V: the family of edges between switching nodes of X:
$$V \subset X \times X.$$
(edges in the control network are also referred to as **control edges**).

We will assume that in G, the control network, the number of edges between a switching node and a terminal node is at most one in each direction.

Given the fact that we have defined a hierarchy function over nodes, it naturally follows that we can decompose the network into layers:

Definition 10.6
A **network layer** of order n is a semi-bipartite subgraph $L_n = ((X_n, Y_n, W_n), V_n)$, where:

- X_n are switching nodes of order n;

- Y_n are terminal nodes of order n;

- W_n and V_n are edges between the nodes of order n.

Each network layer within a heterarchical network, when considered in isolation from all other layers, contains one or more disconnected subgraphs called *fragments*. (A graph $G1$ is said to be **disconnected** from another graph $G2$ if there does not exist a path between any of the nodes in $G1$ and those of $G2$.) If we add the edges linking the different network layers in our telecommunication control network, then we require the resulting graph to be strongly connected.

Definition 10.7
A **fragment** is a strongly connected, disconnected subgraph within a network layer.

We will impose a further restriction on the topology of the network:

Definition 10.8
A heterarchical network is said to be **f-strongly connected** if each network layer in it consists of fragments only.

A telecommunication control network G will have the following properties:

- G is a heterarchical network;

- G is f-strongly connected as well as strongly connected;

- the highest-order layer consists of a single fragment only;

The above model merges two possible extremes; a network being a single fragment on the one hand, and a tree on the other.

It is useful to be able to speak about the set of all edges between two nodes or between sets of nodes:

Definition 10.9
Let A and B denote two sets: $A, B \subset X \cup Y$.
Let Z denote the family of edges in $BG(Z = V \cup W)$.
Let x and y be two nodes from $X \cup Y$.
Then (x, A) denotes the family of edges in Z from x to nodes A:
$(x, A) = \{(x, a) \mid a \in A, (x, a) \in Z\}$; node$x$ is the **initial node** of(x, A).
Likewise, (B, y) denotes the family of edges in Z from nodes in B to node y:
$(B, y) = \{(b, y) \mid b \in B, (b, y) \in Z\}$; node y is the **final node** of (B, y).

With reference to *Figure 10.1 (c)*, let $A = \{b, c, d\}$ and $B = \{a, b, c, d, e, f\}$, then it follows from the above definition that $(a, A) = \{(a, b)\}$ and $(B, b) = \{(a, b), (e, b)\}$.

In addition to being able to define families of edges between node sets, it is also useful, given sets of edges, to define the sets of nodes at the extremities of these edges:

Definition 10.10
Let A and B denote sets as in *Definition 10.6*.
The **final-node function F** is a function from edges to nodes:

$$F(x, A) = \{a \mid a \in A, (x, a) \in Z\}.$$

Likewise, the **initial-node function I** gives the initial nodes:

$$I(B, y) = \{b \mid b \in B, (b, y) \in Z\}.$$

We can extend these definitions:

$$F(B, A) = \bigcup_{\forall a \in A} \{a \mid a \in A, (b, a) \in Z\}, \text{ and } I(B, A) = \bigcup_{\forall a \in A} \{b \mid b \in B, (b, a) \in Z\}.$$

In our notation, with A and B sets, (A, B) denotes a pair of sets as well as the family of edges within Z between the nodes of A and B, assuming $A, B \subset X \cup Y$. With the above functions we get $I(A, B) \cup F(A, B) = A \cup B$.

10.3 Service area

We next proceed by defining the notion of *service area*. We will distinguish several types of them, the most important being the *extended service area*. Let c be a node. The set of all terminal nodes connected to c (with edges from c to these nodes) is the **terminal node service area** $sy(c)$:

(k3) *terminal node service area* :: $sy(c) = \{t \mid t \in Y, (c, t) \in W\}$.

From this definition it follows that $sy(c) = F(c, Y)$. By definition: $sy(\varnothing) = \varnothing$. The terminal node service area of c, together with the set of lower-order nodes with edges pointing from c to these nodes, constitutes the **direct service area** $sd(c)$:

(k4) *direct service area* :: $sd(c) = \{sy(c) \bigcup F(c, X^-(c)),$

where $X^-(c) = \{x \mid x \in X, (c, x) \in V, h(x) < h(c)\}$. By definition: $sd(\varnothing) = \varnothing$. We will also refer to a node in $sd(c)$ as a **descendant** of c. The direct service area relation is additive with respect to a set of nodes:

$$sd(S) = \bigcup_{\forall x \in S} sd(x)$$

Taking the transitive closure of $sd(c)$ - this is the repeated application of the direct service area relation - yields the **service area** $s(c)$:

(k5) *service area* :: $s(c) = sd(c) \bigcup sd \circ sd(c) \bigcup sd \circ sd \circ sd(c) \bigcup \ldots = \overline{sd}(c)$.

The symbol "$\circ$" means function composition. We also assume the network to be finite.

(k6) *fragment* ::
Let *frg(c)* denote the set of nodes within the fragment containing the node c. If S is a set of nodes, then *frg(S)* is the union of the fragments of each element of S:

$$\mathrm{frg}(S) = \bigcup_{\forall x \in S} \mathrm{frg}(x)$$

The operator "frg" extends a set of nodes to include all other nodes within the same fragments. When this operator is applied to the direct service area one obtains:

$$
\begin{aligned}
\mathrm{frg} \circ \mathrm{sd}(c) \quad &= \mathrm{frg}(F(c,Y) \cup F(c,X^-(c))) \\
&= \mathrm{frg} \circ F(c,Y) \cup \mathrm{frg} \circ F(c,X^-(c)) \\
&= F(c,Y) \cup \mathrm{frg} \circ F(c,X^-(c))
\end{aligned}
$$

(k7) *extended service area* ::

Let $FD(c)$ be the short-hand for $frg \circ sd(c)$. Then we obtain the **extended service area** $se(c)$ by applying FD repeatedly:

$$se(c) = c \bigcup FD(c) \bigcup FD \circ FD(c) \bigcup \dots = c \bigcup \overline{FD}(c)$$

Some properties:

1. $c \in se(c)$ reflexivity
2. $b \in se(a) \wedge a \in se(b) \Rightarrow a = b$ (anti-symmetry)
3. $a \in se(b) \wedge b \in se(c) \Rightarrow a \in se(c)$ (transitivity)

Figure 10.2 illustrates the extended service area.

We have obtained knowledge attribute **(k2)** describing network topologies, and attributes **(k3)-(k7)** describing service areas. Next we need to derive three node sets. The next section gives the decision for these sets, as well as the derivation of the required routing predicates.

10.4   The derivation of routing predicates

The decision will be labelled with the name of the specification; since the specification for the next step is the implementation of the previous step (yielding **(k2)-(k7)**), we will use **(k2)-(k7)** as the name:

decision **(k2)-(k7)** ::

The extended service area will serve to determine a path from a source node s to a destination node d. If the path has been set up from the source terminal node s up to node c (we call node c the *current node*), then it has to be determined whether the next path direction will be downward in the hierarchy to a lower-order node or horizontal (i.e. to a node within the same fragment), or upwards in the hierarchy to a higher-order node. For each direction a criterion is needed. These criteria will be derived from three node sets; these node sets will be used to indicate the position of a node within the network at decreasing levels of accuracy. The first node set is the current node:

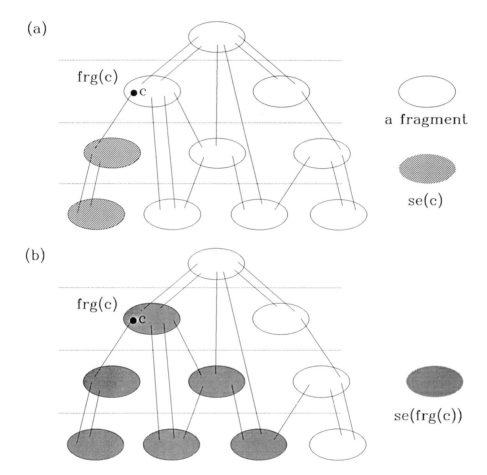

Figure 10.2: (a) the extended service area of node c; (b) the extended service area of the fragment of c.

$\{c\}$

The second set contains the nodes in the same fragment as c, excluding c:

$\mathrm{frg}(c) - \{c\}$

The third set contains all nodes in X *(definition 10.1)*, excluding those that are in $frg(c)$:

$X - \mathrm{frg}(c)$

Notice that $\{c\} \cup (frg(c) - \{c\}) \cup (X - frg(c)) = X$, i.e. the union of the above sets is the set X itself. Also notice that these three sets are disjoint.

Exercise 10.1. Show that $se(X) = X \bigcup Y$, and $se(X \bigcup Y) = X \bigcup Y$.   ∎

Taking the extended service area of the above sets yields:

(k8) *three node sets* ::

$se(c) \qquad se(\mathrm{frg}(c) - \{c\}) \qquad se(X - \mathrm{frg}(c))$

Whereas the first three sets are disjoint, taking the extended service area yields three other sets which may be overlapping (since nodes may be reachable in different ways). If, for example, a node x is both an element of $se(c)$ and an element of $se(frg(c) - \{c\})$, then x is not only reachable via c (and going downward in the hierarchy), but x is also reachable via another node within frg(c). The union of the latter three sets yields $X \cup Y$.

Exercise 10.2. Show that $se(se(S)) = se(S)$, where S is a set of nodes.   ∎

The choice of the three node sets is given by the following argument. If it is assumed that (the majority of) terminal nodes are always in the lowest order network layer, then it is worthwhile going down in the hierarchy as soon as possible; in order to do so one needs information concerning the location of a destination node d. If we know that the destination is in the extended service of c then we know that d is in the third set $se(X - frg(c))$. Therefore the three predicates (and consequently the three node sets) can be seen to give information concerning the location of a node at three levels of refinement.

Using the previously defined node sets the following routing predicates are introduced:

(k9) *routing predicates* ::

$$\begin{array}{lll} P1 & : & d \in se(c) \\ P2 & : & d \in se(\mathrm{frg}(c) - \{c\}) \\ P3 & : & d \in se(X - \mathrm{frg}(c)) \end{array}$$

Predicate $P1$ is true if d is a descendant of c. Predicate $P2$ is true if d is a descendant of at least one of the other nodes in the same fragment. Finally, predicate *P3* states that *d* is a descendant of at least one node outside the fragment.

Current node has a relative meaning in the sense that, although there is always one current node at a time, different nodes can be the current node at different moments in time during the connecting process. In order to refer to a particular routing predicate in a particular switching node, we will write $Pi(c)$, with $i \in \{0, 1, 2, 3\}$, to denote predicate Pi in node c. From the fact that the union of the three sets yields the set $X \cup Y$, it can easily be seen that P1 $\vee$ P2 $\vee$ P3 is always true. In some circumstances, $(P1 \wedge P2 \wedge P3)$ may also be true, because the three node sets are not disjunct in general.

10.5   The connection model m1

These predicates can be used as guards in an alternative construct. The alternative construct has been chosen because the order in which the truth of the predicates is checked is not (yet) of relevance. The checking order will be determined later and is subject to certain design decisions.

With each guard there corresponds a certain action ("command"). More than one guard (predicate) can be true. In that case, however, the command of only one guard that has been found true will be executed, after which the construct terminates. As a consequence, commands corresponding to other true guards are left unexecuted. The following simple alternative construct presents a generic structure for routing in a heterarchical network:

(2) *result of formalization step* ::

```
m0::  if P1   →   dwn(s,d,c)
      ⫾ P2   →   hor⁻(s,d,c)
      ⫾ P3   →   upw(s,d,c)
      fi
```

The commands *dwn()* ("downward") and *upw()* ("upward") reflect the hierarchical character of the communication network, while *hor⁻()* is applied whenever non-hierarchical parts (fragments) are involved. A formal specification of these commands will be given later; an informal description is as follows.

(k10) *routing commands* ::

The command *dwn(s,d,c)* contains the routing strategy in the case where *d* is an element of the extended service area of *c*. The path direction is downward, i.e. towards lower-order network nodes.

The command *upw(s,d,c)* contains the strategy in the case where *d* is not an element of *se(frg(c))*. The path direction is upward (towards higher-order nodes).

The command *hor⁻(s,d,c)* contains the strategy in the case where *d* is within *se(frg(c))* but not within *se(c)*. The path direction is horizontal (towards nodes in the same fragment) until a node is found for which $dwn()$ can be applied (this explains the "$-$" in hor^-). For the alternative construct to terminate properly, at least one guard must be true initially, which is guaranteed by the definition of the routing predicates.

(d2) *decision for the second step* ::

Once the truth of one of the predicates is established, some appropriate actions must follow. Actions on control edges can only be carried out if these edges are present. Therefore, apart from establishing the truth of either $P1$, $P2$, or $P3$, the presence of edges under non-error condition also has to be inspected (the situation in the case of error conditions, i.e. the failure of certain edges or nodes, will be considered in later sections).

Consider the following Supply Predicates $S1$, $S2$ and $S3$:

(**k11**) *supply predicates* ::

S1: there is at least one control edge from c towards a lower-order node x, for which $P1(frg(x))$ is true or $x = d$.

S2: there is at least one control edge from c towards a node within the same network layer.

S3: there is at least one control edge from c towards a higher-order node.

Henceforth $Sj(c)$, with $j \in \{1, 2, 3\}$, will denote supply predicate Sj in node c. In the case of a set of nodes $X, Sj(X) = Sj(a) \vee Sj(b) \vee \ldots \vee Sj(n)$, where $X = \{a, b, \ldots, n\}$. Incorporating these supply predicates changes $m0$ into:

(**3**) *(result of second design cycle)* ::

if	P1 $\wedge$ S1	$\rightarrow$ command 1
▯	P1 $\wedge(\neg$ S1$)$	$\rightarrow$ command 2
▯	P2 $\wedge$ S2	$\rightarrow$ command 3
▯	P2 $\wedge(\neg$ S2$)$	$\rightarrow$ command 4
▯	P3 $\wedge$ S3	$\rightarrow$ command 5
▯	P3 $\wedge(\neg$ S3$)$	$\rightarrow$ command 6
fi		

(**d3**) *decision for third step* ::

However, we can optimize the above result, when making use of some properties of supply predicates.

(**k12**) *properties of supply predicates* ::

If $P1(c)$ is true, then from the definition of the extended service area it follows that there must be a path from c to d via a lower-order node x (possibly: $x = d$). Therefore:

P1 $\Rightarrow$ S1

Similarly, if $P2(c)$ is true, then by definition - because $frg(c)$ is strongly connected - there must be a path from c to d via a node x of the same order. Therefore:

P2 $\Rightarrow$ S2

Similar reasoning for $S3$ does not hold, i.e. if $P3(c)$ is true, then there may not be an edge towards a higher-order node *(Figure 10.3 illustrates this)*.

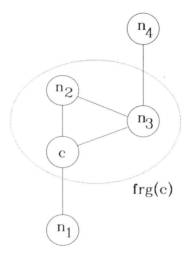

Figure 10.3: Node c in a fragment may not be connected to a higher-order node although $P3$ may be true.

These observations lead to the following updated version of the alternative construct, which will be referred to as $m1$:

(4) *result of third step*

m1::   **if** P1 $\rightarrow$ dwn(s,d,c)
       $\parallel$ P2 $\rightarrow$ hor$^-$(s,d,c)
       $\parallel$ P3 $\wedge$ S3 $\rightarrow$ upw(s,d,c)
       $\parallel$ P3 $\wedge$ ($\neg$ S3) $\rightarrow$ hor$^+$(s,d,c)   (Note: $\neg$ S3 $\Rightarrow$ S2)
       fi

The action $hor^+()$ is similar to $hor^-()$. The path direction is horizontal, but now the aim is to direct the path upwards in the hierarchy as soon as a node within the fragment of c is found for which an edge towards a higher-order node exists.

In summary, using the extended service area of a node or a fragment, three node sets have been defined. This has led to the formulation of three routing predicates. In conjunction with supply predicates (expressing the availability of edges) an alternative construct using guarded commands has been derived. This construct provides a framework for routing in heterarchical networks with distributed control. Four commands have been introduced; they will be refined in later sections.

10.6   A specification of upw, dwn, hor

The following is a specification of *dwn, upw, hor⁻* and *hor⁺*. The specifications are given using pre- and post-conditions. The command $dwn(s,d,c)$ is only executed if $P1(c)$ is true; therefore $P1(c)$ is the pre-condition for $dwn(s,d,c)$. The command $dwn(s,d,c)$ results in the selection of a next node n from the direct service area of c; this is justified because from the definition of extended service area it follows that there is at least one path from c to d going downward.

Apart from selecting a next node, the current node c also realizes a routing relation

   (p,c) R (c,n)

from the incoming edge (p, c) (originating at the previous node p) to an outgoing edge (c, n) towards the next node thta has been selected. An R function is defined as a mapping between $E(X \cup Y, c)$ (the set of edges between $X \cup Y$ and the current node c) and the power set of the edges $E(c, X \cup Y)$ (i.e. the set of subsets of nodes between c and the nodes $X \cup Y$). We are now ready to write the post-condition for $dwn(s,d,c)$.

Let $EST((p,c)\ R\ (c,n))$ express the fact that this R relation has been realized.

{ P1(c) }	pre-condition
dwn(s,d,c)	command
{ n $\in$ sd(c)	post-condition
$\wedge$ (n=d $\vee$ P1(frg(n)))	
$\wedge$ EST((p,c) R (c,n))}	

In a similar way the specification of *upw* is obtained:

$\{S3(c) \wedge P3(c)\}$
   upw(s,d,c)
$\{n \in X^+(c) \wedge EST((p, c)R(c, n))\}$

where $X^+(c) = \{x \mid x \in X, (c, x) \in V, h(x) > h(c)\}$. A relaxed version of the latter specification would be obtained if $P3$ were to be removed from the pre-condition, which would mean that a node can send a connection request to any upward successor.

Specification of *hor⁻* :

{ P2(c) }
  hor$^-$(s,d,c)
{n ∈ frg(c) ∧ EST((p, c)R(c, n))}

Specification of hor$^+$:

{P3(c) ∧ ¬S3(c) ∧ S3(frg(c))}
  hor$^+$(s,d,c)
{n ∈ frg(c) ∧ EST((p, c)R(c, n))}

10.7 The connection model m2

The previous section contained a non-deterministic algorithm $m1$, describing routing in heterarchical networks. The non-deterministic character was used to express that the order in which predicates were checked was of no importance.

(**d4**) *decision to use deterministic model* ::

The model $m1$ serves as a specification of a class of deterministic models in which the checking order has been fixed; these models can be seen as implementations of the non-deterministic specification, although they are still abstract. Let $m2$ denote this class of deterministic models. In this section on example of a deterministic model will be treated.

(**k13**) *deterministic assumption* ::

A deterministic model can be obtained by imposing certain constraints on the checking order of predicates. Determination of this order is guided by the observation that a certain preference exists with respect to it; this preference follows from the following premises:

> PREMISE 1: there is a cost connected with the checking of predicates and therefore as few predicates as possible should be checked;

> PREMISE 2: if it is assumed that terminal nodes are in the lowest-order network layer, then a path should go downward in the network hierarchy as soon as possible.

From this it follows that the preferred checking order is: $P1$, $P2$, $P3$. In case where the premises 1 and 2 do not hold, the preferential checking order should be reconsidered. However it does not influence the method of working adopted in this chapter.

From (P1 ∨ P2 ∨ P3) the following implication can be inferred:

(**k14**) :: ¬ (P1 ∨ P2) ⇒ P3

This, together with the previously obtained preferred checking order, gives the following algorithm, belonging to $m2$:

(5) *result of the fourth step* ::

ALGORITHM A (order: P1, P2)
   if   P1   **then**   dwn(s,d,c)
   elseif P2   **then**   hor⁻(s,d,c)
   elseif S3   **then**   upw(s,d,c)
                else   hor⁺(s,d,c)
   fi

If all fragments only contain one node (i.e. a fragment is a node), then algorithm A reduces to the following structure:

if $d \in s(c)$ then $dwn(s, d, c)$ else $upw(s, d, c)$ **fi**,

where $s(c)$ is the service area of c.

10.8   Error predicates - their effect on m1

(d5.1) *first decision for the fifth step* ::

Ideally, our domain of interest consists of f-strongly connected heterarchical networks, i.e. networks for which all disconnected subgraphs within the network layers are strongly connected. In practice, however, edges may be removed and nodes may not be able to perform their switching function due to failure, congestion, etc. The actual network may therefore differ from the ideal switching network.

(k16) *availability of network elements* ::

Let X.av, Y.av, W.av and V.av denote the available nodes and edges of the control network G. Hence, G.av is the control network consisting of its available elements.

(k17) ::

The error predicates will be derived according to the structure of the routing predicates $P1$, $P2$ and $P3$. The first error predicate $E1$ is derived from $P1$ in the following way. First it should be noted that the truth of $P1$ in a node c implies that there is a node n in $sd(c)$ for which the following holds:

$$(n=d) \vee d \in se(frg(n))$$

This in turn implies that both n and the control circuit from c to n are available.

(k18) *control circuit* ::

Let C denote a function relating an input control edge with an outgoing control edge between the same nodes e.g. $(c, n) = C(n, c)$. The pair $< (c, n), (n, c) >$ related in this way constitutes a **control circuit**. The direction of a control circuit is determined by the direction in which a connection request is sent.

Let VW.av = V.av∪ W.av and let XY.av = X.av∪ Y.av. Then the error predicates can be stated as:

(k19) *error predicates*

$E1(c) :: \forall n : n \in sd(c) \wedge ((n = d) \vee d \in se(frg(n))) :$
$\quad \neg((c, n) \in VW.av \wedge C^{-1}.(c, n) \subset VW.av \wedge n \in XY.av)$
$E2(c) :: P2(c) \wedge (\forall n : n \in F.E(c, frg(c)) :$
$\quad \neg((c, n) \in V.av \wedge C^{-1}.(c, n) \subset V.av \wedge n \in X.av))$
$E3(c) :: P3(c) \wedge S3(c) \wedge (\forall n : n \in F.E(c, X^+(c)) :$
$\quad \neg((c, n) \in V.av \wedge C^{-1}.(c, n) \subset V.av \wedge n \in X.av))$

Using these error predicates, the generic connection model $m1$ can be reformulated. This results in a model which is at a similar abstraction level as the original $m1$. However, an additional factor is taken into account, namely the fact that nodes and edges may not be available.

(k15) *decision to use repetitive construct* ::

Notice that if one of these error predicates is true, the normal termination of the alternative construct may not be possible. In that case one of the other commands should be executed if its corresponding guard is found true. This, of course, is only applicable in cases where the node sets overlap, i.e. when alternative paths exist. In those cases more than one guard may be true. To this end, the alternative construct will be replaced by a repetitive construct The repetitive construct reflects the possibility of a re-trial (within the same node) in case a certain outgoing edge or successor node is not available.

Because of its repetitive character, termination of the construct should be guaranteed. Dijkstra [Dij76] formulates a variant function for this purpose. If the number of true guards is taken as the variant function, then this number is bound to decrease if a construct like: **do B→ B:=false od** is used.

(d5.2) *second decision for the fifth step* ::

In order to incorporate the effect of the error predicates, the commands *dwn(s,d,c)*, *upw(s,d,c)*, *hor⁻ (s,d,c)* and *hor⁺ (s,d,c)* will be replaced by the boolean operators *DWN(s,d,c)*, *UPW(s,d,c)*, *HOR⁻ (s,d,c)*and *HOR⁺ (s,d,c)*; they are defined in the same way as the previous commands, but in addition they return the value *true* or *false*. A value *false* indicates that the corresponding error predicate is *true*, i.e.:

$E1 = \neg DWN(s, d, c), E2 = \neg HOR^-(s, c, d) = \neg HOR^+(s, d, c)$ and
$E3 = \neg UPW(s, d, c).$

The reason for taking the negation is that a value *true* for *DWN(s,d,c)* may be interpreted as its successful termination (and hence, $E1$ must have been false).

With B initially true and $E1$, $E2$, and $E3$ initially false, a possible refinement of $m1$ is:

(6) *result of fifth step*

```
m1::do   P1 ∧¬E1 ∧ B          → E1:= ¬ DWN(s,d,c); B:=E1
     ▯   P2 ∧¬E2 ∧ B          → E2:= ¬ HOR⁻(s,d,c); B:=E2
     ▯   P3 ∧ S3 ∧¬E3 ∧ B     → E3:= ¬ UPW(s,d,c); B:=E3
     ▯   P3 ∧ (¬S3) ∧¬E2 ∧ B  → E2:= ¬ HOR⁺(s,d,c); B:=E2
   od;
```

When this construct terminates with B *true*, then an exception condition arises and appropriate action should take place (like sending an error message).

In this repetitive construct, the truth of an error predicate results in making certain guards false (i.e. they are switched off). However, instead of switching a guard off as soon as an error predicate is found true, this action could be postponed until the error predicate has been found *true* a number of times; (in this way, setting up a connection could be retried several times). Hence, one could introduce a counter function, which would count the number of times a particular error predicate is found *true*. Then the corresponding guard would only be switched off if the counter obtained a certain value.

10.9   Error predicates - their effect on m2

(d6) *decision of the sixth step* ::

Using the checking order $P1$, $P2$, $P3$, a sequential implementation of the previous result can be obtained. This yields the following result:

(7) ALGORITHM A (order: P1, P2) ::

```
while B do            (B initially true)
if      P1 ∧¬E1       then E1:= ¬DWN(s,d,c); B:=E1
elseif  P2 ∧¬E2       then E2:= ¬HOR⁻(s,d,c);B:=E2
elseif  P3 ∧ S3 ∧¬E3  then E3:= ¬UPW(s,d,c); B:=E3
elseif  P3 ∧¬S3 ∧¬E2  then E2:= ¬HOR⁺(s,d,c);B:=E2
else "error message"; B:= false
fi
od;
```

Chapter 11

THE PROCESS MODEL

11.1   Introduction

This section combines and refines the results of *Sections 9* and *10*. The synthesis of the model to de derived here (the *PROCESS MODEL*) will be largely based on informal reasoning, followed by formal verification. In order to carry out this design cycle, we have to indicate its associated decision, its specification, and its knowledge attribute.

The general line along which the implementation will proceed is as follows. We will use the connection model from the previous section as the knowledge attribute for the current step; hence, we interpret this knowledge attribute as the *internal model* of the system we are designing (whereas the communication model specifies the *external behaviour*). We will use *algorithm A* of *Section 10.9*.

We will then decompose the two models (*communication model* and *connection model*) into sets of communicating agents; this requires that the designer creates a decomposition and defines the behaviours of each agent as well as the way they are interconnected.

The communication model will be implemented as a process, the *RESPONSE PROCESS* (RES). *Algorithm A* of *Section 10.9* will be refined into a number of processes, the *IDENTIFICATION* and *SELECTION PROCESSES*. The ensemble of these processes will be referred to as the *COMPOSITE; Figure 11.1* illustrates how the constituent processes of *COMPOSITE* have been obtained from *algorithm A*. The response process *RES* implements the communication behaviour of a switching node with the external world, whereas *COMPOSITE* reflects the internal structure of a switching node with respect to the connecting function. The process *COMPOSITE* will deal exclusively with connection requests; consequently we need another process to discriminate between connection requests and other messages. This process will be referred to as ENT (from *ENTRANCE PROCESS*).

In summary, *ENT* will receive connection requests and sends these to *COM-*

171

POSITE; in addition, it sends connection requests as well as secondary control information to *RES*; (**secondary control information** is any control message other than a connection request). The relationship between these processes and *algorithm A* is shown in the following picture:

ALGORITHM A (*E*1, *E*2, and *E*3 are initially false; *B* is initially true)

   while B do

   if   ⌜ $\overline{P1}\ \overline{\wedge\neg E1}$ ⌝   **then**   ⌜ $\overline{E1} := \overline{\neg DWN(s,d,c)};$ ⎤
              | | | B := E1 |
              | | | |
              | PID | | |
              ⌞ _ _ _ ⌟   ⌞ _ _ _ _ _ _ _ _ _ DWN _⌟

   elseif ⌜ $\overline{P2}\ \overline{\wedge\neg E2}$ ⌝ **then**   ⌜ $\overline{E2} := \overline{\neg\ HOR^-(s,d,c)};$ ⎤
              | | | B := E2 |
              | | | |
              | SID | | |
              ⌞ _ _ _ ⌟   ⌞ _ _ _ _ _ _ _ _ _ HOR⁻ _⌟

   elseif P3 ∧ S3 ∧¬E3   **then** | $\overline{E3} := \overline{\neg\ UPW(s,d,c)};$
                                         | B := E3 |
                                         ⌞ _ _ _ _ _ _ _ _ _ _ UPW _⌟

   elseif P3 ∧¬S3 ∧¬E2 **then**   ⌜ $\overline{E2} := \overline{\neg\ HOR^+(s,d,c)};$ ⎤
                                         | B := E2 |
                                         ⌞ _ _ _ _ _ _ _ _ _ HOR⁺ _⌟

   else "error message"; B := **false**

   fi
   end;

Figure 11.1 shows all the constituent agents after the decomposition. During the formal specification of each of the agents, also the ports via which they communicate have to be defined. Hence, *Figure 11.2* is the refined of the block diagram of *Figure 11.1*

This model is based on the assumption of *S*3 being true; consequently HOR⁻ is not present (in case *S*3 is true, the process UPW has to be replaced by the process HOR⁺). For notational convenience, only the integers of port names have been given; the remaining parts can be easily derived. For example, the input port to RES, labelled 1, is the port *res.in1* etc. Expressions between round brackets () indicate predicate values. Because a switching node may itself be implemented as a network, this model is essentially a hereditary one, i.e. it applies to nodes, groups of nodes, hierarchies of nodes, etc.

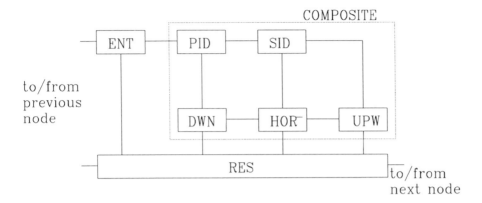

Figure 11.1: (a) *Algorithm A* as initial model for *COMPOSITE*; dotted lines indicate processes *PID, SID, DWN, HOR⁻,* and *HOR⁺,* and *UPW.* (b) architectural entities of a switching node, where each entity carries one of the processes.

Blocking behaviour is due to the possible unavailability of edges and nodes. This may be caused by failures in cables, etc. Another reason is the following: since processes will be mapped onto physical media they will consequently have a finite processing capacity. The limited capacity will be assumed to be revealed in the following ways:

a. Processes have a finite processing speed.

b. Message channels between processes have zero buffer capacity (one could include these using the method of *Section 4.2.*

c. Processes have a finite number of message channels (i.e. ports).

The connecting process within a node may receive many control messages, generally from many different directions. Because of its limited capacity, the connecting process can only accept control messages at a limited rate (the *call rate*). The limited capacity will give rise to a loss of connection requests if the rate at which they are being offered is too high. Therefore, the connecting process has to use part of its capacity to control the use of its remaining capacity. This function will be incorporated into *ENT*, the *entrance process.*

11.2 The entrance process

Function: 1) to receive control information and discriminate between connection requests and secundary control information;

2) to check the free capacity of subsequent processes.

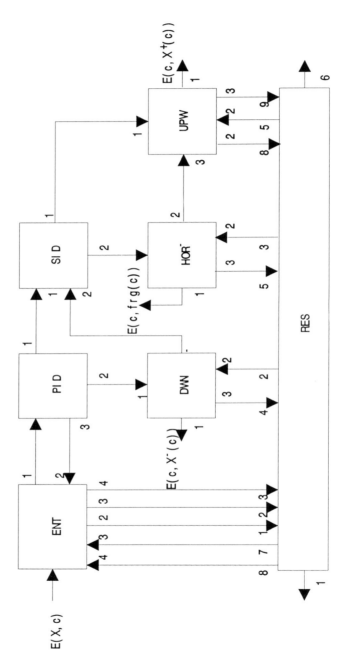

Figure 11.2: Interconnection scheme of the Process model of a switching node.

3) to route connection requests to the primary identification process *PID* if that process has sufficient processing capacity.

4) to route connection requests to the response process *RES* if sufficient capacity for PID does not become available within a certain time (in which case we will speak of a *non-granted connection request*). If the capacity of the *response process* is not sufficient to accept a non-granted connection request, then this request will be deleted.

5) to route secondary control information directly to the response process *RES*.

From this it follows that only connection requests may be blocked; secondary control information will always be accepted. In order to show the influence of the finite processing capacity of the processes *PID* and *RES*, the specification of *ENT* will be done in two steps.   Let $\overline{con}$ denote any control message except a connection request (e.g. $\overline{con}$ = ptr). Let $ent.in?\overline{con}$ be a short-hand notation for:

ent.in? ster + ent.in? ptr + ent.in? ntr + ent.in? dter.

Then the specification of *ENT* is:

E1 = ent.in1?$\overline{con}$: E1	(*ENT* neglects any control info as long as)
+ ent.in1?con E2	(no connection request has been received)
E2 = ent.out4!con:	(it sends a copy of the connection request)
ent.out1!con: E3	(to *RES* and another copy to *PID*)
E3 = ent.in3?ster : E1	(the entrance process is reset by a source)
	(termination request from *RES*)
+ent.in1?$\overline{con}$: E4	(the entrance process is transparent to)
	(all secondary control information;)
E4 = ent.out3!$\overline{con}$: E3	(a message received, while in E3, is sent)
	(to the response process)
+ ent.in3? ster:E1	

Let *pid* and *res* denote the remaining processing capacity (in number of connection request) of *PID* and *RES* respectively. Incorporating the finite processing capacity of RES and PID into the equations for ENT yields:

E1 = ent.in1?$\overline{con}$: E1	
+ ent.in1? con :	(after receiving a connection request, *ENT*)
ent.in2?*pid* :	(inputs the values of the remaining)
ent.in4?*res* : E2	(processing capacities of *RES* and *PID*)

E2 = **if** pid > 0 **and** res > 0 **then** ent.out4! con: ent.out1!con : E3
    + **if** ¬ pid > 0 **and** res > 0 **then** ent.out2! con : E3
    + **if** res ≤ 0 **then** E1

$$E3 = \text{ent.in1?}\overline{\text{con}} \quad : E4$$
$$\quad\quad + \text{ent.in3?ster} \quad : E1$$

$$E4 = \text{ent.out3!}\overline{\text{con}} \quad : E3$$
$$\quad\quad + \text{ent.in3?ster} \quad : E1$$

11.3 Identification processes

Identification processes check the different routing predicates $P1$, $P2$ and $P3$, i.e. they inspect whether the destination node is within certain extended service areas.

The primary identification process - PID

Function: to determine whether the destination node is within the extended service area of the current node. If $d \in se(c)$ then the connection request is sent to the downward selection process DWN. If $\neg(d \in se(c))$ then the connection request is sent to the secondary identification process SID.

Let pid denote the remaining capacity of the PID process (i.e. the number of connection requests it still can process), then the specification of PID is:

$$P1 = \quad \text{pid.in? con} : P2$$
$$\quad + \text{pid.out3! pid} : P1 \quad\quad \text{(value of remaining capacity is input by } ENT)$$

$$P2 = \quad \text{if} \quad d \in se(c) \text{ then pid.out2! con} : P1$$
$$\quad + \text{if} \neg \ (d \in se(c)) \text{ then pid.out1! con} : P1$$

The secondary identification process - SID

Function: to determine whether $d \in se(frg(c) - \{c\})$. If so, the connection request is sent to the *horizontal selection process HOR$^-$*, otherwise it is sent to the *upward selection process UPW*, or (if the latter has not been implemented) to the alternative horizontal selection process HOR$^+$.

The specification for SID is:

$$S1 = \quad \text{sid.in1? con} : S2$$
$$\quad + \text{sid.in2? con} : S2 \quad\quad \text{(connection request received from DWN}$$
$$\quad\quad\quad\quad\quad\quad\quad\quad\quad\quad\quad\quad\quad\quad \text{because error predicate E1 was found}$$
$$\quad\quad\quad\quad\quad\quad\quad\quad\quad\quad\quad\quad\quad\quad \text{true)}$$

$$S2 = \quad \text{if } d \in se(fgr(c) - \{c\}) \text{ then sid.out2!con} : S1$$
$$\quad + \text{if } \neg(d \in se(frg)(c) - \{c\})) \text{ then sid.out1!con} : S1$$

11.4 Selection processes

This class of processes contains the downward, the horizontal and the upward selection processes. The name is chosen because within these processes, edges are selected between nodes on a path between a source node (terminal node) and a destination (terminal) node. In *Section 10.6* the *R function* has been introduced. Selection processes establish *R* functions, and implement respectively the commands *dwn(s,d,c)*, *hor(s,d,c)* and *upw(s,d,c)* of *algorithm A*.

The downward selection process - DWN

Function: 1. Upon receipt of a connection request, DWN selects an available outgoing edge to an available node within the direct service area and sends the connection request via this edge. It also sends a copy of the connection request to the response process.

2. If no available edge or node is present (error predicate $E1$) it sends the connection request to the secondary identification process.

The DWN process implements the *dwn(s,d,c)* command. It contains the downward routing strategy and relates incoming edges with outgoing edges. It realizes the *R* function for downward directions; this will be indicated as the *dwnR* function, with the following domain and range:

$$\text{dwnR: } E(X,c) \rightarrow \mathcal{P} \left(E(c,X^-(c)) \right) \qquad (\mathcal{P} \text{ is power set})$$

This definition covers *distribution* of a call. However, in almost all cases, connections yield a one-to-one correspondence:

$$\text{dwnR : } E(X,c) \rightarrow E\left(c,X^-(c)\right)$$

The *dwnR* function assigns a subset of $E(c, X^-(c))$ to an element of $E(X, c)$. From the above expression it follows that:

$$\text{dwnR.}E(X,c) \subset E(c,X^-(c))$$

Let (con, R) denote a message consisting of a connection request and R. The value of the R function applied to the incoming edge via which the request entered the switching node; (R returns an outgoing edge). Then the specification of the downward selection process is as follows:

D1 = dwn.in1?con : D2   (after receipt of a connection)
                                                   (request, *DWN* performs the)
D2 = **if** DWN(s,d,c) **then** dwn.out3!(con,R):D1   (command *dwn(s,d,c)*; if $E1$ is)
                                                   (found false, the connection)
                                                   (request, together with the)

(value of the R function)

(is sent to RES;)

(otherwise DWN signals the)

(failure of the selection)

$+$ **if** $\neg$DWN(s,d,c) **then** dwn.out2! con:D1

The horizontal selection process - HOR$^-$

Function: 1. Upon receipt of a connection request this process selects an available outgoing edge to an available node within the same fragment and sends a connection request via this outgoing edge. It also sends a copy of the connection request to the response process.

2. If no available edge or node is present (error predicate $E2$) it sends the connection request to UPW if $E(c, X(c))$.

This process implements the $hor^-(s,d,c)$ command. It contains the strategy for non-hierarchical routing within fragments. This process is only present if $frg(c)\neq c$. An alternative view is that HOR^- implements the $dwn()$ command on fragment scale, i.e. it implements $dwn(s,d,frg(c))$.

The path direction is horizontal and nodes within the fragment apply their horizontal selection processes until a node p within the fragment is reached for which $d \in se(p)$, upon which the path direction is downward in the network hierarchy. The alternative horizontal selection process HOR^+ is similar to HOR^- except that in this case the path direction is upward in the network hierarchy; HOR^+ is used whenever a node p within the fragment is found for which there is no upward outgoing edge, i.e. $E(p, X^+(p)) = \varnothing$. An alternative view is that HOR^+ implements $upw(s,d,frg(c))$. The specification for HOR^- is as follows:

H1 = hor.in1? con : H2

H2 = **if** HOR$^-$(s,d,c) **then** hor.out3! (con,R): H1
$+$ **if**$\neg$ HOR$^-$(s,d,c) **then** hor.out2! con : H1

The specification of HOR$^+$ can be obtained in a similar way.

The upward selection process - UPW

Function: 1. Upon receipt of a connection request this process selects an available outgoing edge to an available node in a higher-order selected outgoing edge. It also sends a copy of the connection request to the response process.

2. If no available edge or node is present it sends the connection request to the response process.

This process implements the $upw(s,d,c)$ command. It contains the strategy to be followed if the destination node is not within the extended service area of the fragment containing the current node.

The specification for the upward selection process:

U1 = upw.in1? con : U2   (*UPW* has two inputs for connection)
    + upw.in3? con : U2   (requests; one from *SID* and one)
                                                  (from HOR$^-$)

U2 = **if** UPW(s,d,c) **then** upw.out3! (con,R): U1
    + **if** ¬ UPW(s,d,c) **then** upw.out2! con : U1

11.5   The response process

Function: 1. To receive from *ENT* non-granted connection requests.

2. To receive connection requests from selection processes (after *R* functions have been established).

3. To receive secondary control information from *ENT* and to take care of the necessary actions instructed by that information.

4. To receive non-granted connection requests from selection processes.

The response process generates and receives secondary control information. It also receives non-granted connection requests. As has been said in the introduction to this chapter, the response process implements the communication model of the switching node, whereas the identification and selection processes represent the connection model.

Starting with the communication model of a switching node *C* (equations **(18)** in *Section 9.6)*, we will derive a model for the response process by performing three refinement steps.

The *first step* is based on the assumption that a switching node will not show internal delay (notice that equations **(18)** were derived under the assumption that internal delay will occur within a network. This new assumption will yield simpler equations. Recalculating the behaviour equations for a switching node under the assumption that no delay occurs inside a switching node (but blocking may still occur), we obtain equations *C'*:

C0 =	c.in1? con	: C0a		C5 =	c.in1? ster	: C7
					+c.out1! ptr	: C8
C0a =	τ : C0b + τ : C1				+c.in2? dter	: C6
C0b =	c.in1? ster	: C0		C6 =	c.in1? ster	: C10
	+c.out1! ntr	: C0c			+c.out1! ntr	: C11
C0c =	c.in1? ster	: C0		C7 =	c.out2! ster	: C0
	+c.in2? dter	: C10				
C1 =	c.out2! con	: C3				
	+c.in1? ster	: C0		C8 =	c.in1? ster	: C7

$$+c.in2? \text{ dter } : C9$$

$$
\begin{array}{llll}
C3 = & c.in1? \text{ ster} & : C4 & \\
     & +c.in2? \text{ ptr} & : C5 & \quad C9 = \quad c.in1? \text{ ster} \quad : C10 \\
     & +c.in2? \text{ ntr} & : C6 & \qquad\qquad\; +c.out1! \text{ dter} : C11
\end{array}
$$

$$
\begin{array}{llll}
C4 = & c.out2! \text{ ster} & : C0 & \quad C10 = \quad c.out2! \text{ ster} \quad : C0 \\
     & +c.in2? \text{ ptr} & : C7 & \\
     & +c.in2? \text{ ntr} & : C10 & \quad C11 = \quad c.in1? \text{ ster} \quad : C10
\end{array}
$$

Also, the behaviour of a switching node without internal delay is embedded in the behaviour of a switching node with delay (this is a property of the combination algorithm); this is the subject of the next exercise..

Exercise 11.1 Show that $C(C', C') \approx C$ ∎

The *second step* is to replace this set of equations by a set of equations which is observation congruent with the previous set. This involves two steps, during which additional τ's are introduced to model internal actions. First consider the term:

$$C0a = \tau : C0b + \tau : C1$$

According to the τ-4 laws, the right-hand side is observation congruent with: $\tau : C0b + \tau : (\tau : C0b + \tau : C1)$. This form will be used as it expresses that the response process has, in fact, two blocking behaviours. The first is the result of finite processing capacity of the switching node and is given by the first $\tau : C0b$ term. However, the response process can also show blocking behaviour because the selection processes did not find an available outgoing edge; in that case the connection request cannot be sent due to the failure of an outgoing edge. This second blocking behaviour is modeled by the second $\tau : C0b$ term.

According to the first τ-1 law, $C0a$ can be replaced by $\tau : C0a$ in equation $C0$:

$$
\begin{aligned}
\tau : C0a = \; & \tau : (\, \tau : C0b + \tau : C1) \\
= \; & \tau : C0b + \tau : (\tau : C0b + \tau : C1) & (\tau\text{-4}) \\
= \; & \tau : C0b + \tau : (c.in1?ster : C0 + \tau : C0b + \tau : C1) & (\tau\text{-4}) \\
= \; & \tau : C0b + \tau : C1' \, ,
\end{aligned}
$$

where $C1' = c.in1?ster : C0 + \tau : C0b + \tau : C1$.

Next, insert τ's after *c.in1? ster* in $C0b$, $C0c$, $C1$ and $C1'$; insert τ's after *c.out2! ster* in $C4$, $C7$ and $C10$. Finally, replace $C0$ by $\tau : C0$ to the right-hand side of $C0$. These τ's will reflect the internal communication of *RES* with the other processes. The resulting equations after this second refinement step are:

$$
\begin{array}{llll}
C0 = & c.in1? \text{ con} & : C0a' & \quad C5 = \quad c.in1? \text{ ster} \quad : C7 \\
     & +c.out1! \text{ ptr} & : C8 &
\end{array}
$$

$$C0a' = \quad \tau : C0b + \tau : C1'$$

$$
\begin{aligned}
C0b = \quad & c.in1? \text{ ster} \quad : \tau : C0 \\
+ & c.out1! \text{ ntr} \quad : C0c
\end{aligned}
$$

$$
\begin{aligned}
C0c = \quad & c.in1? \text{ ster} \quad : \tau : C0 \\
+ & c.in2? \text{ dter} \quad : C10
\end{aligned}
$$

$$
\begin{aligned}
C1 = \quad & c.out2! \text{ con} \quad : C3 \\
+ & c.in1? \text{ ster} \quad : \tau : C0
\end{aligned}
$$

$$
\begin{aligned}
C1' = \quad & c.in1? \text{ster} \quad : \tau : C0 \\
+ & \tau : C1 \\
+ & \tau : C0b
\end{aligned}
$$

$$
\begin{aligned}
C3 = \quad & c.in1? \text{ster} \quad : C4 \\
+ & c.in2? \text{ ptr} \quad : C5 \\
+ & c.in2? \text{ ntr} \quad : C6
\end{aligned}
$$

$$
\begin{aligned}
C4 = \quad & c.out2! \text{ ster} \quad : \tau : C0 \\
+ & c.in2? \text{ptr} \quad : C7 \\
+ & c.in2? \text{ntr} \quad : C10
\end{aligned}
$$

$$+c.in2? \text{dter} \quad : C6$$

$$
\begin{aligned}
C6 = \quad & c.in1? \text{ ster} : C10 \\
+ & c.out1! \text{ ntr} : C11
\end{aligned}
$$

$$C7 = \quad c.out2! \text{ster} : \tau : C0$$

$$
\begin{aligned}
C8 = \quad & c.in1? \text{ ster} : C7 \\
+ & c.in2? \text{ dter} : C9
\end{aligned}
$$

$$
\begin{aligned}
C9 = \quad & c.in1? \text{ster} \quad : C10 \\
+ & c.out1! \text{dter} : C11
\end{aligned}
$$

$$C10 = \quad c.out2! \text{ster} : \tau : C0$$

$$C11 = \quad c.in1? \text{ ster} : C10$$

In the *third step*, the communication with the other processes is fully implemented. However, this last step will not be shown in all detail.

First of all, a number of relabellings are performed:

Replace:	by:
c.in1? ster	res.in2? ster
c.out1! ptr	res.out1! ptr
c.out1! ntr	res.out1! ntr
c.out1! dter	res.out1! dter
c.out2! ster	res.out6! ster
c.in2? ptr	res.in2? ptr
c.in2? ntr	res.in2? ntr
c.in2? dter	res.in2? dter

With respect to connection requests, the picture has to be refined still further. First of all, the response process can receive connection requests via two input ports; res.in1 and res.in3 respectively. The first is used for non-granted connection requests. The second is used for granted requests. Hence, instead of:

$$C0 \; = c.in1? \text{ con} : C0a$$
$$C0a = \tau : C0b + \tau : C1'$$

we can write:

$C0 = res.in1?\ con\ :\ C0b + res.in3?\ con:\ C1'$

In $C1$, a connection request is output at $c.out2$.

The above three refinement steps yield the following specification of the response process; the message (con,R) is the connection request and the routing function R, needed to find the required outgoing edge).

$$
\begin{aligned}
R0\ \ =\ &res.in1?\ con\ \ \ :\ R1 &&(\text{non-granted connection request; })\\
+\ &res.in3?\ con\ :\ R2 &&(\text{granted connection request; })\\
+\ &res.out8!res\ \ :\ R0 &&(\text{capacity output for } ENT\ )
\end{aligned}
$$

$$
\begin{aligned}
R1\ =\ \ \ &res.in2?\ ster\ :\ res.out7!\ ster\ :R0 &&(\text{blocking behaviour })\\
+&res.out1!ntr\ :res.in2?ster:\ res.out7!ster\ :\ R0 &&(\text{non-granted })\\
&&&(\text{requests })
\end{aligned}
$$

$$
\begin{aligned}
R2\ \ =\ &res.in2?ster\ :\ R4\\
+\ &res.in4?(con,R)\ :\ R3\\
+\ &res.in5?(con,R)\ :\ R3 &&(\text{These terms correspond to } \tau\ :\ C1\ )\\
+\ &res.in7?(con,R)\ :\ R3 &&(\text{in the previous section })\\
+\ &res.in9?(con,R)\ :\ R3\\
+\ &res.in6?con\ :R1 &&(\text{These terms correspond to } \tau\ :\ C0b\ )\\
+\ &res.in8?con\ :R1 &&(\text{in the previous section; they })\\
+\ &res.in10?con\ :\ R1 &&(\text{reflect blocking behaviour for })\\
&&&(\text{granted connection requests})
\end{aligned}
$$

$$
\begin{aligned}
R3\ \ =\ &res.in2?ster:\ res.out7!\ ster\ :\ R0\\
+\ &res.out6!con\ :\ R5\\
R4\ \ =\ &res.in4?(con,R):\ res.out7!\ ster:\ R0\\
+\ &res.in5?(con,R)\ :\ res.out7!\ ster\ :\ R0\\
+\ &res.in7?(con,R)\ :\ res.out7!\ ster\ :\ R0\\
+\ &res.in9?(con,R)\ :\ res.out7!\ ster\ :\ R0\\
+\ &res.in6?con\ :\ res.out7!\ ster\ :\ R0\\
+\ &res.in8?con\ :\ res.out7!\ ster\ :\ R0\\
+\ &res.in10?con\ :\ res.out7!\ ster\ :\ R0
\end{aligned}
$$

$$
\begin{aligned}
R5\ =\ \ \ &res.in2?ster\ \ \ :\ R6\\
+\ &res.in2?ptr\ \ \ \ :\ R7\\
+\ &res.in2?ntr\ \ \ \ :\ R8
\end{aligned}
$$

$$
\begin{aligned}
R6\ =\ \ \ &res.out6!ster\ :\ res.out7!\ ster\ :\ R0\\
+\ &res.in2?ptr\ \ \ \ :\ R9\\
+\ &res.in2?ntr\ \ \ \ :\ R12
\end{aligned}
$$

$$
R7\ =\ \ \ res.in2?ster\ \ \ :\ R9
$$

$$
\begin{array}{lll}
& + \text{ res.in2?dter} & : \text{ R8} \\
& + \text{ res.out1!ptr} & : \text{ R10}
\end{array}
$$

$$
\begin{array}{lll}
\text{R8} = & \text{res.in2?ster} & : \text{ R12} \\
& + \text{ res.out1!ntr} & : \text{ R13}
\end{array}
$$

$$
\begin{array}{lll}
\text{R9} = & \text{res.out6!ster} & : \text{ res.out7! ster} : \text{R0} \\
& + \text{ res.in2?dter} & : \text{ R12}
\end{array}
$$

$$
\begin{array}{lll}
\text{R10} = & \text{res.in2?ster} & : \text{ R9} \\
& + \text{ res.in2?dter} & : \text{ R11}
\end{array}
$$

$$
\begin{array}{lll}
\text{R11} = & \text{res.in2?ster} & : \text{ R12} \\
& + \text{ res.out1!dter} & : \text{ R13}
\end{array}
$$

$$
\begin{array}{lll}
\text{R12} = & \text{res.out6!ster} & : \text{ res.out7! ster} : \text{R0} \\
\text{R13} = & \text{res.in2?ster} & : \text{ R12}
\end{array}
$$

11.6   Verification of the result

Verification with respect to the connection model

The processes *PID, SID, DWN, HOR⁻ , HOR⁺* , and *UPW* implement the connection model $m2$. We calculate, using the expansion law:

COMPOSITE = (PID|SID|DWN|HOR⁻ |UPW){p1,p2,p3,p4,p5,p6} where

$$
\begin{array}{llll}
\text{p1} = & \text{pid.out1} & = & \text{sid.in1;} \\
\text{p2} = & \text{pid.out2} & = & \text{dwn.in1;} \\
\text{p3} = & \text{sid.out1} & = & \text{upw.in1;} \\
\text{p4} = & \text{sid.out2} & = & \text{hor.in1;} \\
\text{p5} = & \text{dwn.out2} & = & \text{sid.in2;} \\
\text{p6} = & \text{hor.out2} & = & \text{upw.in2}
\end{array}
$$

The composition can be easily performed because of the inherent sequential nature of this part of the process model (due to the sequential character of *algorithm A*). This simplifies the evaluation of the expansion theorem. Let $P1S1D1H1U1$ denote the initial state of the ensemble of identification and selection processes; this state is constructed from the initial states of the constituent processes (i.e. $P1$, $S1$, $D1$, $H1$ and $U1$ respectively). Starting in this global state, subsequent states can be derived by expanding the equations for *PID, SID, DWN, HOR,* and *UPW*. This gives:

$$
\begin{aligned}
\text{P1S1D1H1U1} = &\text{ pid.in? con : P2S1D1H1U1 + pid.out3! pid : P1S1D1H1U1} \\
\text{P2S1D1H1U1} = &\text{ if d} \in \text{se(c) then } \tau \text{ : P1S1D2H1U1} \\
& + \text{ if } \neg \text{ (d} \in \text{se(c)) then } \tau \text{ : P1S2D1H1U1}
\end{aligned}
$$

P1S1D2H1U1 = if DWN(s,d,c) then dwn.out3! (con,R): P1S1D1H1U1
                    + if ¬ DWN(s,d,c) then τ : P1S2D1H1U1
P1S2D1H1U1 = if d ∈ se(frg(c)−{c}) then τ : P1S1D1H2U1
                    + if ¬ (d ∈ se(frg(c)−{c})) then τ : P1S1D1H1U2
P1S1D1H2U1 = if HOR⁻(s,d,c) then hor.out3! (con,R): P1S1D1H1U1
                    + if ¬ HOR⁻(s,d,c) then τ : P1S1D1H1U2
P1S1D1H1U2 = if UPW(s,d,c) then upw.out3! (con,R): P1S1D1H1U1
                    + if ¬ UPW(s,d,c) then upw.out2! con : P1S1D1H1U1

Using the τ-1 law, and renaming the behaviour identifiers, these equations can
be written as:

COMPOSITE::

   P1 =   pid.in? con:P2 + pid.out3! pid : P1

   P2 =   if d ∈ se(c) then D2
        +if¬(d ∈ se(c)) then S2

   D2 =   if DWN(s,d,c) then dwn.out3! (con,R): P1
        +if¬DWN(s,d,c) then S2

   S2 =   if d ∈ se(frg(c)−{c}) then H2
        +if¬(d ∈ se(frg(c)−{c})) then U2

   H2 =   if HOR⁻ (s,d,c) then hor.out3! (con,R): P1
        +if¬HOR⁻ (s,d,c) then U2

   U2 =   if UPW(s,d,c) then upw.out3! (con,R): P1
        +if UPW(s,d,c) then upw.out2! con : P1

In the following step, the communication with the outside world is omitted. This
can be modeled by replacing all communication terms by τ . Applying the τ-1
law, and using the definition of observation equivalence yields the following set
of equations:

COMPOSITE.m2::

   P1 =   τ : P2

   P2 =   if d ∈ se(c) then D2
        +if¬ (d ∈ se(c)) then S2

   D2 =   if DWN(s,d,c) then P1
        +if¬ DWN(s,d,c) then S2

   S2 =   if d ∈ se(frg(c)−{c}) then H2
        +if¬ (d ∈ se(frg(c)−{c})) then U2

   H2 =   if HOR⁻(s,d,c) then P1

+if¬ HOR⁻(s,d,c) **then** U2

U2 =   **if**   UPW(s,d,c) **then** P1
         +**if**¬ UPW(s,d,c) **then** P1

It is now easy to recognize connection model $m2$ (algorithm A) in these equations by replacing "if... + if..." by "if... ‖ fi".

Verification with respect to the communication model

First, the communication behaviour of *COMPOSITE* will be considered, then the combined behaviour of *COMPOSITE* and *RES* will be calculated. We obtain the communication behaviour of *COMPOSITE* by replacing the conditions in the conditional behaviour terms by τ's (if-reduction).

P1 =   pid.in?con : P2 + pid.out3! pid : P1
P2 = τ : D2 + τ : S2
D2 = τ : dwn.out3! (con,R) : P1 + τ : S2
S2 = τ : H2 + τ : U2
H2 = τ : hor.out3! (con,R) : P1 + τ : U2
U2 = τ : upw.out3! (con,R) : P1 + τ : upw.out2! con : P1

Applying the τ-4 law to $P2$ and $S2$, and substituting these terms into $P1$ and $D2$ respectively, yields:

P1 =   pid.in? con : D2 + pid.out3! n : P1
D2 = τ : dwn.out3! (con,R) : P1 + τ : H2
H2 = τ : hor.out3! (con,R) : P1 + τ : U2
U2 = τ : upw.out3! (con,R) : P1 + τ : upw.out2! con : P1

Substituting $U2$ into $H2$, and $H2$ into $D2$ yields:

P1 =   pid.in? con : D2 + pid.out3! pid : P1
D2 = τ : dwn.out3! (con,R) : P1
         +τ : (τ : hor.out3! (con,R) : P1
             + τ : (τ : upw.out3! (con,R) : P1
                 τ + : upw.out2! con : P1))

By using the following identities: hor.out3 = upw.out3 = dwn.out3 = comp.out3; we obtain the following set of equations for COMPOSITE:

P1 =   pid.in? con : D2 + pid.out3! pid : P1
D2 = τ : comp.out3! (con,R) : P1
         +τ : (τ : comp.out3! (con,R) : P1
             + τ : (τ : comp.out3! (con,R) : P1
                 + τ : upw.out2! con : P1))

Finally, $D2$ can be reduced using τ-4 and the τ-1 law into:

$D2 = \tau : (\tau : \text{comp.out3! (con,R)}: P1 + \tau : \text{upw.out2! con} :P1)$

The first τ can be removed according to the τ-1 law, since $D2$ is used after a guard in $P1$. The resulting equations are:

COMPOSITE.com::

$P1 = \text{pid.in? con} : D2 + \text{pid.out3! pid}: P1$

$D2 = \tau : \text{comp.out3! (con,R)}: P1 + \tau : \text{upw.out2! con} : P1$

Again using the expansion theorem, together with the CCS laws, we will next calculate and reduce the following expression:

$(\text{ENT}|\text{PID}|\text{SID}|\text{DWN}|\text{HOR}|\text{UPW}|\text{RES})\backslash P = (\text{ENT}|\text{COMPOSITE}|\text{RES})\backslash P$

where P is the set of connections between the different processes. As we are interested in the communication behaviour of the process model, it will suffice to use *COMPOSITE.com*. Hence:

$(\text{ENT}|\text{COMPOSITE.com}|\text{RES})\backslash P$

The equations for the response process can be simplified by the following relabelling:

Replace:	by:
res.in4	comp.out3
res.in5	comp.out3
res.in7	comp.out3
res.in9	comp.out3
res.in6	comp.out2
res.in8	comp.out2
res.in10	comp.out2

This yields identical terms, which can be removed using the absorption law for summation. As *ENT* is transparent to all secondary control information, we are able to use simplified model for *ENT*:

ENT' :: E1 = ent.in1? con: ent.in2? n : ent.in4? m: E2
        E2 = τ : ent.out2! con : ent.in3? ster : E1
           + τ : ent.out4! con : ent.out1! con : ent.in3? ster : E1

Hence, only connection requests are dealt with in this model for *ENT*. All secondary control information is assumed to be sent to *RES* directly.

From *Figure 11.2* it follows that: ent.out1 = pid.in; ent.out2 = res.in1; ent.out4 = res.in3; ent.in2 = pid.out3 and ent.in4 = res.out8.

Consider the composition (ENT'|COMPOSITE.com|RES)\ P, with :

P = { res.in1,res.in3,res.out8,pid.out3,pid.in,res.out7,comp.out3,comp.out2 }
Using the following relabelling:

Replace:	by:
res.in2? ster	c.in1? ster
res.in2? ptr	c.in2? ptr
res.in2? ntr	c.in2? ntr
res.in2? dter	c.in2? dter
res.out1! ptr	c.out1! ptr
res.out1! ntr	c.out1! ntr
res.out1! dter	c.out1! dter
res.out6! con	c.out2! con
res.out6! ster	c.out2! ster

we find, using the expansion theorem, after reduction using the CCS-laws:

(ENT' |COMPOSITE.com|RES)\ P = C'

Exercise 11.2. Derive the meta program for this chapter.   ■

11.7 Data model

We have obtained a detailed model showing the internal and external communications of a switching node in the context of call processing. In this model we have focussed on the *synchronization framework (Section 2.7)*. We can now use this to add operations on data.

First of all, one can interpret actions to consist of port names and message values. Second, by defining abstract data types, one can specify operations such as an *SDL-task*; within CCS expressions, such tasks would have the semantic value τ. Access functions specified within the abstract data type can be used to build port names. For example, given a data structure D consisting of several fields, one of which would be *D.field1*, one can build the action *out!D.field1* to send the value of that data field.

Another possibility would be to use the functions of the abstract dat types to build port names; this is required in case one wants to designate input or output ports which is required in switching applications. For instance *D.field2!D.field1* means that a message, the value of which can be found in *D.field1* is sent via an output port, the name of which is determined by *D.field2*. See also [OP90]. *LOTOS* [Gel87] is based on CCS and combines communication and data structures.

Part III

DESIGN, CREATIVITY AND LEARNING

Chapter 12

DESIGN ITERATIONS

12.1 Iterations and the basic design cycle

Iterations in the design process are needed to support the learning process in-volved. The basic design cycle enables a concise description of the design process with the objective to manage it effectively.

The sequence of design cycles that a designer carries out shows repetitions of some (or maybe all) design cycles. This is due to the fact that during a design cycle the designer may find that the resulting implementation does not satisfy the specification. Hence, iterations of design cycles occur due to the fact that the designer learns while designing. In most (if not all) cases, recording this sequence of design cycles is not very suited as a means of capturing the design process and representing that process afterwards. The learning process which took place should be reflected in the updating of knowledge attributes, and not in repeating the iterations. Hence, the design process should be rewritten into a representation that can be better comprehended for later re-use or that can be used to communicate with other designers.

In *Chapter 8*, we made a distinction between the *actual design process* (as it has been carried out by the designer), and the *ideal design process* (as it will be represented afterwards). Let us consider for a moment the type of iterations in both the ideal and the actual design processes.

The occurrence of iterations is the subject of [Dow86, Koo86] and is implicit in the basic design cycle. Iterations reflect the effectivity of the *actual design process*; if a designer has full knowledge of the steps of be taken, then the design process can proceed without iterations. Hence, iterations and the actual design process are closely linked. This is in contrast with the *ideal design process*. There, hierarchical design descriptions using the basic design cycle, in the way we have explained in *Chapters 8-10*, are not meant to model iterations, but to manage the complexity of the design process and its associated design objects. Using *DPDL*, for instance, allows one to discern design processes and their objects at different

191

levels of detail, while preserving the basic structure of the design process.

We have treated in great detail the use of formal techniques to design communicating systems. We have focused on the modeling of ideal design processes using *DPDL* in our examples of *Chapters 9-11*. Let us now turn our attention more to the actual design process, to the role of learning and human intelligence in it, and to the use of computers. In this chapter we will look at the occurrence of iterations from several perspectives: we will consider the impact of Gödel's work. We will also consider the design process from the viewpoints of chaos theory and the computability of designs, followed by a view based on the entropy of design.

Chapter 13 will then look at the consequences of what we have learned in this chapter. In particular, we will focus on the aspect of learning as well as on the consequences of iterations and learning on CAD and knowledge representation.

From the existence of iterations in actual design processes, but also from the point of view of complexity management in ideal design processes, the question emerges as to whether one can measure the complexity of a design cycle. This stems from the expectation that more errors will occur when the steps taken in a design cycle are too large, and the designer has to redo such a step several times, leading to iterations. Therefore in *Chapter 14* we will develop a design metrics, which will provide us with a measuring rod by which we can measure the structural complexity of implementations against their specifications. The method of measurement is based on the definition of language vocabularies, from which the implementations are constructed. The vocabulary spans a universe of solutions; hence, the solution that is chosen is one of potentially many, and this represents the information which corresponds with the obtained implementation.

12.2   Gödel's theorem

In any consistent formal system one can write true statements which cannot be deduced from the axioms and rules of the formal system.

The work of the mathematician Kurt Gödel has had a tremendous impact on our thinking about the essence of mathematical reasoning. Gödel showed that any consistent formal system containing the natural numbers is fundamentally incomplete; i.e. there are true statements which cannot be deduced from the axioms and rules of the formal system. We will consider the impact of this result on design. The interested reader is referred to [NN58], which contains a more extensive treatment of the Gödel results with more references.

Gödel showed how to represent *meta-mathematical reasoning*, i.e. how to represent statements and conclusions *about* mathematical formulas, as formulas *in* a formal system. In this case he used the formal system of the positive numbers as the formal system; a meta-mathematical reasoning is thereby mapped onto a formula relating positive numbers.

Gödel showed that it is possible to associate in a unique way a number to either a symbol, a formula or a proof. Such a *Gödel number* can be used to translate meta-mathematical statement about expressions and their relationships into a statement about their corresponding Gödel numbers and their mathematical relations. Gödel derived a formula G (which we will not show in detail). Let $GN(G)$ be the Gödel number of G. Gödel showed that G is the image within the calculus of the meta-mathematical statement

"the formula with Godel number $GN(G)$ cannot be proved".

From this it follows that the formula G within the calculus corresponds with the meta-mathematical statement

"the formula 'G' cannot be proved".

Hence, formula G within the calculus states about itself that it cannot be proved.

Gödel next showed that *if* formula G can be proved, than also its negation can be proved and vice versa. But, if both a formula and its negation can be proved (i.e. derived from the axioms of the formal system), then the axioms of the formal system are inconsistent. Stated otherwise:

If the axioms are consistent then G is undecidable.

Since *(i)* true meta-mathematical statements are translated into valid formulas within the calculus, *(ii)* the translation is one-to-one (due to the way Gödel numbers are constructed), and *(iii)* the meta-mathematical reasoning concerning G was true,

we must conclude that G is a valid formula within the calculus.

We have observed that: *(i)* G is undecidable, and *(ii)* G is true. From these observations we conclude that **the axioms are incomplete.** Moreover, they are fundamentally incomplete; adding more axioms (provided they remain consistent) could make the previously undecidable formula decidable, but then another true, but undecidable formula could still be defined.

Hence, we are forced to acknowledge a fundamental weakness in any formal system. It is impossible to define a set of axioms that would enable one to derive from these axioms all conceivable true statements. However, it is possible to add new axioms from which previously unprovable statements could be proved. However, there would be other true statements which could not be derived from the new axioms, etc.

Let us now return to our issue of iterations in the design process and try to reformulate Gödel's result in that context. During the design process, a designer will certainly use the power of reason in order to arrive at the best possible solution to a design problem. If we accept the view that this type of reasoning is a formal system in the sense of Gödel, and if we translate *"true statement in the calculus"* by *'solution to a design problem'*, then we conclude that there may be solutions to a design problem which cannot be found purely by a formal reasoning process.

Seen from this perspective, we may speculate that the iterations are a designer's way to discover new axioms, from which better design solutions could be found using a formal reasoning. We all share the feeling of sudden awareness that "the pieces are falling together". It is as though the designer breaks out of a formal system and at the same time creates another one, until a logical line of reasoning is established leading to a required solution. In this way, iterations during design and Gödel's result share a fundamental relationship and are therefore intimately linked together.

12.3   Design and computability

Not all functions are computable; i.e. there are not enough computer programs to compute every possible function.

After Gödel's result, mathematicians became interested in provability rather than truth. The mathematician Hilbert had posed a number of fundamental problems. His 23-rd problem was: *discover a method for establishing the truth or falsity of any statement in the predicate calculus.*

Alan M. Turing discovered that Hilbert's problem was impossible to solve. He did this by defining more precisely the concept of computation. Beginning with the idea that a computation is an algorithm - a procedure that can be carried out mechanically without creative intervention - he showed how this idea can be refined into a detailed model of the process of computation in which any algorithm is broken down into a sequence of simple, atomic steps. The resulting model of computation is the logical construct called a *Turing machine* [Hop84].

Without going into detail about the nature of a Turing machine it can be stated that: given a large but finite amount of time, the Turing machine is capable of any computation that can be done by any modern digital computer, no matter how powerful. Although any modern computer can work many times faster than a Turing machine, the Turing machine has become indispensable for the theoretical study of the ultimate problem-solving capacity of any real computer. One could ask the question whether any function could be realized by a Turing machine. The answer to this question is negative and goes as follows. A Turing machine can read and change the contents of a tape (the *program*). The function that a particular Turing machine can perform can be expressed as the string of characters on its tape. Hence, any Turing machine can be expressed as a character string of finite length. Therefore, it is possible to list all possible Turing machines in numerical or alphabetical order. Then one can pair this ordered list one for one with the set of whole numbers. There is no fixed upper limit to the size of the Turing machine, so there is no limit to the number of Turing machines. Hence, the set of all possible computable functions is the same size as the set of all whole numbers (both sets are called countable sets).

However, it can be shown that there are infinite sets that are not countable; they are larger than the set of whole numbers in the sense that they cannot be paired one for one with the whole numbers. One example of such a non-countable set is the set of all the functions of the positive whole numbers that take on integer values. A careful analysis shows that there must be more such functions than there are whole numbers. The implication is that not all functions are computable; there are not enough computer programs to compute every possible function.

One could ask the question whether a human is capable of computing every possible function (provided the person would live long enough) or whether the person is subject to the same rule.

Within the class of computable functions, one can discern problems that are tractable (i.e. which can be solved within a reasonable amount of time), and problems which take an unrealistic amount of time. This is the subject of mathematical complexity theory, where two important categories of problems are recognized: P-complete and NP-complete problems [GJ79].

First of all, we say that a problem is complete (for a class of problems) if the following statement holds: *there is an efficient way of solving the problem only if there is an efficient way of solving any other problem in the class.*

A problem is designated to be a $P-$problem if it can be solved in polynomial time, that is: it's solution is some polynomial function of the size of the problem. For example, a binary sort is an example of a problem which can be solved in polynomial time (in this case it depends on the logarithm of the number of elements to sort). A problem is assigned to class P only if no instance of that problem requires more than polynomial time to solve. The class P is important since many computer scientists regard problems that are not polynomial as intractable.

One can also define a non-deterministic Turing machine; it is allowed to solve a problem by guessing the answer and then verifying the guess. For example, to determine whether an integer is composite, the non-deterministic machine guesses a divisor, divides and, if the division is exact, verifies that the number is composite. The deterministic machine must search systematically for a divisor.

The time needed to solve a problem with a non-deterministic machine is measured by the length of the shortest computations, and so it would seem that the non-deterministic machine has a great advantage over the deterministic one. Ordinary experience suggests it is easier to verify a solution than it is to find it in the first place. However, the non-deterministic machine may take a long time before it finds this shortest computation.

The class of problems designated NP, is the class of problems that can be solved with a non-deterministic machine in polynomial time. For example, the problem of colouring a graph in such a way that no two nodes connected by an edge have the same colour is known to be NP-complete. No one has yet been able to prove that NP-problems are intrinsically more difficult to solve than the problems in the class P. Whether or not the class P is distinct from the class NP (this is called the $P - NP$ problem), has become one of the major open questions in modern mathematical complexity theory.

An important question is whether P and NP problems are equivalent or not. It appears that this question cannot be settled at the moment. There are computations for which one can assume either that P and NP are equivalent or not, without detriment to the consistency of the system. This fact is generally taken as evidence that solutions to such problems are beyond the current state of mathematics.

Again, we will bring the above reasoning within the context of design iterations. If no iterations occur, then the designer is able to find a solution on the basis of the available design knowledge and building blocks. In Turing's terms, the designer acts more like a deterministic "machine" in that case.

However, if not enough design knowledge is available, the designer may try to guess a solution and verify the result, in which case the designer's behaviour reflects that of a non-deterministic Turing machine. As soon as the problem is more fully understood, the designer can put this design knowledge into a systematic framework and define a line of reasoning for solving the design problem.

It looks like Gödel and Turing consider two sides of the same coin. However, the result of Gödel's reasoning as we discussed it earlier is not identical with the concept of a non-deterministic Turing machine. The latter is non-deterministic but fixed (in the sense that the choices have been defined before), whereas finding new axioms in the Gödel sense is a non-deterministic but probably dynamic activity. The distinction between the deterministic and non-deterministic worlds is not always obvious as we learn from the theory of chaos.

12.4   Design and chaos

Chaos is a phenomenon that can be studied separately. Unlike its name suggests, there is some order in the mechanisms underlying chaotic behaviour in physical processes.

In their article, Crutchfield et al. [CFPS86] describe how chaos imposes fundamental limits on prediction. They argue that microscopic causes can lead to macroscopic effects. They base their reasoning on the theory of dynamic systems; a dynamic system consists of two parts: the notions of a state (the essential information about a system) and a dynamic (a rule that describes how the state evolves with time).

The behaviour of a dynamic system can be represented as an orbit in a state space. Crutchfield et al. use the example of a simple pendulum, of which the state space is constructed with two dimensions; position and velocity. In the case of a friction-less pendulum the orbit is a loop. If the pendulum encounters friction, the orbit spirals to a point as the pendulum comes to rest. Such a point is called an *attractor*. The set of points in state space that evolve to an attractor is called its basin of attraction. A pendulum clock has two such basins: small displacements of the pendulum from its rest position result in a return to rest; with large displacements however, the clock begins to tick as the pendulum executes a stable oscillation (with the help of a mainspring or weight to overcome losses of friction). Attractors may also have other shapes than points. In the case of the pendulum, the second stable situation is where the clock ticks at a steady rate and the attractor is a cycle in state space.

Attractors of the above type allow us to predict the behaviour of dynamic systems. In contrast, attractors in the case of chaotic systems do not allow such prediction. Any initial information from which the future behaviour could be predicted, very rapidly looses its predicting value due to the exponentially growing influence of microscopic perturbations. In the case of chaotic systems, the attractors are *fractals*, objects that reveal more detail as they are increasingly magnified.

Crutchfield et al. use a simple stretching and folding operation, which takes place in the state space, to further improve our understanding of chaotic behaviour. One can imagine what happens in the state space model by imagining that a baker puts a drop of blue food colouring in the dough and then kneads it. The kneading is a combination of two actions: rolling out the dough, in which the food colouring is spread out, and folding the dough over. Only after a very limited number of steps (say, 20) the food colouring has been distributed throughout the dough. Another way to illustrate this is by imagining a bucket of water. The water rotates, and we put a drop of ink into the water. After only a few rotations of the water, the ink is already distributed.

During its movement, a particle from the food colouring or ink encounters a dough particle or water molecule. At the time of collision, microscopic effects greatly influence the trajectory of the particle. After only a few such collisions, the trajectory of the particle bears very little relationship with its original trajectory. In this case the chaotic attractor is caused by the random behaviour of the dough particles or water molecules. One may observe such random fluctuations by observing the movements of a small dust particle in the air; a random influence is superimposed on the gravitational effect; colliding air molecules result in fluctuating behaviour of the dust particle (this is called its *Brownian movement*).

The stretching and folding of a chaotic attractor systematically removes the initial information and replaces it with new information: the stretch makes small-scale effects larger, the fold erases large scale information. Hence, it becomes impossible to derive the initial state of a system from an observation of its current state, or make a prediction about the future state of a system. The existence of chaos is a fundamental principle.

Returning to the issue of design iterations, the following association may be interesting. We have defined several basic mechanisms in *Section 1.4*. In case of design, the rolling out (as described in the dough example above) can be compared with the operation *decompose*, whereas the operation *compose* can be compared with the folding action. If we accept this view, then the above theory also explains why a design cycle involving "rolling/stretching" operations, when performed by different designers, can yield so many different solutions from the same initial conditions.

The theory of chaos learns us that relatively simple mechanisms may lead to complex or seemingly random behaviour; in the presence of suitable attractors chaotic behaviour can result. The viewpoint that random fluctuations (which lead to chaotic behaviour) may actually be a cause for creative behaviour is a thought that is reflected in the *entropy* view on the design process [Koo85b], which is the subject of the next section.

12.5   Design and entropy

Rules can be formulated which characterize some properties of the design process with respect to its non-deterministic behaviour. This entropy view on the design process can be stated in terms of a quantitative and qualitative description of the involved information flows.

The idea of using information theory to gain a better understanding of systems was used by Conant [Con76], although Brillouin [Bri62] already applied a similar idea with respect to scientific reasoning. Conant used the theory to better understand real world systems. He admits that "*There are obvious dangers in applying information theory, designed for use under the severe mathematical constraints of stationarity and ergodicity, to real-world systems not thus constrained*". However, the justification lies in the fact that instead of "*being content to say nothing about information*", a far more preferable course seems to be to "*try to use results from the formal theory by judicious interpretation and generalization*".

Conant defines the **entropy rate** of a variable X as the entropy of X conditional on all its prior values, which is the information carried per observation in a long sequence.

Next he obtains an expression for the **total information rate** F (in bits /sec) as a measure of the total processing activity within a system:

$$F = F_t + F_b + F_c + F_n \text{ (bits/sec)}$$

The term F_t is the **throughput rate** and is a measure of the relatedness between input and output; it is the term which transmission engineers wish to optimize in the case where the system is a transmission channel.

The second term is the **blockage rate** and represents the effort needed by the system in order to block non-relevant information.

The **coordination rate** F_c represents the amount of information processing needed to obtain a coordinated action among the subsystems. For example, commands from a microprocessor to its memory or to peripheral circuitry connected via a bus are needed for the internal coordination.

The **noise rate** F_n reflects the amount of information in a system that is not reflected in its input. For instance, in a transmission channel, the noise rate reflects the amount of thermal noise produced in that channel.

$H(A/B)$ is the **conditional information** in A, given B; it is the amount of information in A when B is known. For instance, if we throw a dice with outcome '2' then $H(2) = 1/6$; however, if we know that the outcome is less than three (let us call this K), then $H(2/K) = 1/2$.

The noise rate can be expressed as H (system/input), i.e. the information in the system, conditional on the input. Or, the information in the system when its input is known. This term is zero if the output is determined by the input (e.g. the system is a multiplier in which case the output is determined by a function which computes its value as a product of the inputs).

$T(A : B)$ is the **transmission** between A and B and is a measure of the relatedness between A and B. The following qualities hold:

$$T(A:B) = H(A) - H(A/B) = H(B) - H(B/A).$$

The transmission is zero if A and B are independent and maximum if one determines the other. In the case of a noiseless transmission channel, then $T(A : B) = H(A) = H(B)$. We will often use the following decomposition rule:

$$H(A,B) = H(A) + H(B/A) = H(B) + H(A/B).$$

where $H(A, B)$ is the information corresponding to the combination of A and B. In the sequel we will not consider the information carried per observation in a long sequence, but only the information per observation (i.e. the dimension of the expressions will be "its" rather than "bits/second").

Suppose a designer D is given a specification S and produces and implementation I. Then the term $H(D/S)$ represents the amount of information produced by the

designer, given the specification S. We expect this term to be positive. If not, all
the information to produce the implementation would already be contained in
the specification. This non-zero term corresponds to the noise term in Conant's
expression for the total information rate.

We can decompose the above term on the basis of the following assumption.
During the design cycle from S to I, the designer produces the mental concepts
or design language which is used to describe I as well as the design knowledge
K which is needed to derive I from S, such that I is consistent with S (i.e.
the implementation I is correct with respect to the specification S). Hence,
$D = (K, L, I)$ in our model; using the above decomposition rule we get:

$$H(D/S) = H(K,L,I/S) = H(K,L/S) + H(I/K,L,S)$$

The first term is the information in the creation of the design language L and
the design knowledge K (see also *Figure 1.1*). The creation of L may consist
of the selection of an already existing language from a set of possible languages.
For example, the design language may be the library of standard cells used in
the design of an integrated circuit. The designer may also add new concepts,
like user defined types or procedures in a programming language. However, the
designer might equally well construct a completely new language - a situation
which occurs, for instance, when a designer expresses a design in terms of block
diagrams.

The second term in the expression $H(D/S)$ is the information in I in dependence on the language and design knowledge used and the model from which
I was derived. If K is such that it enables the designer to derive I from S,
then by definition the amount of information in the implementation, given the
specification and the knowledge, is zero. This is expressed as the

First information law for system design (*FILSD*):

$$H(I/K,L,S) = 0$$

Expression (*FILSD*) also implies that if K can be expressed in some formal language (for example, a computer program), then the transformation from S into
I can be obtained using a deterministic machine if the latter is given the design
knowledge K in a formalized manner, e.g. as an algorithm. As long as the
knowledge K is insufficient and hence does not satisfy (*FILSD*), the required
implementation I cannot be obtained. Expression (*FILSD*) is the analogon of
the first law of thermodynamics (conservation of energy) in the domain of information creation during design processes.

It seems reasonable to assume that the creative and intellectual capabilities of
a designer are not infinite. Considered as an information processing system, the
designer has a finite information processing capacity. Let $C_d(S, t))$ denote the

amount of information which a designer can produce within a time t with a view to finding an implementation of S. Obviously, each designer has a different level of experience and background, and therefore a designer has a unique capacity function. Suppose a designer has a time t available for the implementation step from S to I. Then the creation of information by the designer is bounded by the above capacity function. This is expressed as the

Second information law for system design (*SILSD*):

$$H(I/S) \leq C_d(S,t)$$

Under the condition that the implementation should be correct with respect to the specification, it is assumed that the capacity function C may not increase very much (or even decrease) beyond a certain complexity (information content). We will call this the *saturation effect*. This is based on the assumption that a very limited number of information items can be contained in a person's "working memory"; if a design would exceed that number then the designer is likely to introduce errors when trying to solve others.

In practice, design or development projects have a target in terms of their throughput time. If the required information exceeds the capacity function at the time a project should be finished, then the implementation I cannot be found without any design errors as there is not yet a sufficient amount of design knowledge K.

Allowing more time would not be the first solution to try for projects that are under time pressure. Also, allowing more time may not solve the problem as well due to the mentioned saturation effect. Another solution is to decrease design complexity by introducing intermediate design levels. In [Koo79] it is shown that the structural information in a design rapidly decreases if one introduces intermediate design levels. This supports the use of a hierarchical design method as we have done in *Chapters 8-11*.

We could also put more people on the project, thereby increasing $C_d(S,t)$. However, this requires an efficient organization as well as skilled people. The effect of putting more people on the job may not always lead to design time reduction as was shown in [Bro75]. Adding computer power could lead to an increase of F_n, as the computer takes over certain tasks from the designer. Other factors may cause F_n to decrease. This will be treated in *Chapter 13*.

It may be worthwhile to make the following observation. Often, people who are trained in theoretical computer science try to put enough information into the specification such that an implementation can be generated from it. Hence, they have some difficulty in grasping the idea of *entropy* (or its counterpart *information*) in a design. However, this can be overcome by realizing that in order to write a specification in this way, the designer would still have to create the specification. An amount of information would then be associated with that specification.

Using the two information laws, we can draw the following conclusions with respect to the occurrence of design iterations. The first law requires that sufficient design know how is available. If not, then a consistent specification cannot be obtained from the specification by applying a line of reasoning based on this knowledge. Therefore, a new iteration is required where new knowledge would be obtained, leading to a situation in which the required implementation can be obtained.

If the design language is fixed, then the required information in the creation of the implementation I is $H(I/L, S)$. Then the number of iterations needed to create I would be:

$$H(I/L,S) \ / \ C_d(S, t_{cycle})$$

where $C_d(S, t_{cycle})$ is the amount of information produced by the designer in time t_{cycle}, the time the designer uses for the synthesis part of a design cycle. If the designer chooses this time too short (trying a "quick fix" and then inspect the result), then many iterations occur. If this time is chosen long enough, then only one cycle would be needed (provided no saturation would occur).

Chapter 13

THE ROLE OF CAD AND LEARNING IN DESIGN

13.1 Computations in CAD

Computers can only do what they are told to do. The capability of using a CAD tool in an interactive way is a fundamental requirement.

In the previous chapter we have discussed the occurrence of design iterations from several viewpoints. In this chapter we will study the consequences of these viewpoints on the role and limitations of Computer Aided Design. We will also discuss the aspect of learning and propose some learning mechanisms in view of the results of the previous chapter. First we will consider the role of computers from the viewpoint of Gödel's results[Koo88].

The axioms and derivation rules of the formal system within a computer are constructed from its gates, memories, busses and software. Following these axioms and derivation rules means that a computation proceeds in a deterministic way.

Based on the above reasoning and on the results of *Section 12.2*, we conclude that a computer in support of the design process cannot derive a true formula which is undecidable. Stated otherwise, a computer cannot do anything creative. The 'braking out of the system' that usually is required during a design process cannot be provided by a computer. A designer however could, in principle, derive formulas that are undecidable within the formal system. Such situations cause design iterations to occur. Hence, this is a strong argument to make any CAD tool interactive in such a way that a designer could add a true formula. Although the CAD tool would not be able to derive such formula, the CAD tool could use it as a valid formula within its formal system from the moment such formula has been entered by the designer.

With reference to the basic design cycle, the interactivity requirement is valid for both the *synthesis* or *construction* phase and for the *evaluation* or *verification* phase. There is a distinct difference between these two phases. The result of the synthesis phase is an implementation of the specification and therefore contains more detailed design information, whereas the evaluation phase is carried out on the basis of the availability of both a specification and an implementation and in this sense no detailed design information has to be added. Since the synthesis phase contains most of the creative part of the design cycle, we may expect CAD for that activity to be more difficult to realize in the sense that it requires the encoding of design knowledge. Interactivity in the case of CAD for evaluation purposes (such as *simulation*) therefore differs from that for construction or synthesis purposes in the sense that the latter require the addition or changing of design rules and building blocks. In both cases, interactivity also means the interactive use of the CAD tool as it is.

Hence, we conclude that the capability of using a CAD tool in an interactive way is a fundamental requirement. We can observe the effect of this requirement in the case of high level synthesis tools, where the designer of the CAD tool may continue to add more design rules in order to enable the CAD tool to cover more design situations and improve its performance.

The discussion about the role of CAD would not be complete without considering the impact of artificial intelligence techniques. In the 1950's when artificial intelligence research started, many people believed that the age of mechanized human capabilities had arrived at last. In many textbooks and magazines one could read about the miraculous things that could be expected in a few years. The initial enthusiasm about artificial intelligence became frustrated for two reasons. One reason was the lack of understanding of 'intelligent behaviour' (such as reasoning, learning, understanding, vision, etc.). The second reason was the total inadequacy of the compute power available in those days.

Let us first view the field of artificial intelligence from the perspective of the required compute power. A human brain contains about a thousand billion neurons, each capable of firing at most ten times a second. This may indicate an upper bound for the total information processing capacity of the brain in the order of ten-thousand billion bits per second. Over the last twenty or thirty years we have seen a speed improvement of a factor of ten in five years. Assuming that this trend would remain valid for the next twenty to thirty years, then we may expect that in that period compute power could reach the same compute power level as the brain.

Although achieving the above speed improvements would itself already be a formidable task, the lack of understanding of intelligent behaviour may be a more limiting factor in our aim to make computers more intelligent. Let us briefly consider the elements of intelligence. One can characterize intelligence by means of the following capabilities:

1. the ability to learn or understand from experience;

2. the ability to respond quickly and successfully to a new situation;

3. the use of the facility of reason in solving problems.

In order to be able to learn from one another's experience, one must capture that experience in a way which makes it suitable for future use. The introduction of *DPDL* in *Chapter 8* is meant for this purpose. *Section 13.3* will consider this aspect further.

The capability to react successfully to unexpected situations is something computers do very poorly; if a situation is expected, it is being programmed into the software, if it is not expected, it is not in the computer program and therefore the computer will not react (let alone successfully); a possibility is to enable a computer to infer from its state and the new input that an abnormal situation has occurred. It may then enter an *exception handling* mode where it will evaluate possible scenario's.

Using the facility of reason in solving problems is something a computer can do very well as long as the formal system as used by the computer is adequate. Although a computer is very weak in the first two capabilities, it may still appear intelligent due to its high speed when performing certain calculations where it performs far better than a human being. Adding more compute power is not sufficient, however, in reaching a true creative behaviour. A discussion of this can also be found in [CC90, Sea90], where one author defends the view that adding more compute power will lead to improved levels of intelligence, whereas the second author refutes this viewpoint.

The great mathematician Alan M. Turing (1912-1954) who has made fundamental contributions to the mathematical understanding of computerized reasoning, has come up with a thought experiment which has come to be known as the *Turing test* to define intelligent behaviour. The experiment goes as follows. Imagine a room with a terminal. A person sits behind this terminal and starts a conversation. The terminal is connected to a piece of equipment in a second room which cannot be observed by this person. The equipment in the second room could be another terminal with a second person sitting behind it, or it is a computer taking part in the conversation. If the first person is unable to distinguish between the two situations, then we conclude that the computer in the second room exhibits intelligent behaviour. When we apply the Turing test to the field of design, then

we could consider a computer intelligent when it finds an implementation which, in terms of design performance, cannot be distinguished from a solution by an expert human designer.

It is an interesting question in how far the $P-NP$ problem is related to the design problem: for instance, CAD tools are capable of dealing with problems within the class P. In case no polynomial time solution is available, expert systems can be used which contain heuristic rules; these are pieces of knowledge associated with non$-P$ problems which have been put into a form which itself is P. In this sense, expert systems are a tool to enable the designer to establish a link between the non$-P$ world and the world of P-problems.

13.2   CAD and entropy

Design involves the creation of information during a learning process. Our information laws for system design provide insight into the way Computer Aided Design is linked to the design process.

This section discusses the impact of our information model on the ways in which computer tools can be used. From the first information law for system design *(FILSD)* we know that a deterministic machine has to be provided with the design knowledge K in order to be able to perform a detailing step. Hence, we see that deterministic machine tools (CAD tools) can only perform a detailing step when an algorithm, a look-up table, a procedure, is available in the machine language. This illustrates the first type of use of machine tools:

(a) as implementation machines using algorithms to perform detailing steps.

Examples of this type of use are machines which design the lay-out and wiring of circuitry or silicon compilation tools to generate logic designs from high level descriptions. The notion of *interactive CAD* summarizes the discussion from the previous chapter. Other tools are not so much aiming at executing or supporting the design cycle activities themselves, but rather in supporting the organization of the design process. Hence these tools aim

(b) to support the capturing and execution of design processes, for example on the basis of *DPDL* or similar approaches.

We would speak of truly intelligent tools if our CAD tools of both categories would be capable of learning and acquiring design knowledge. Let us next consider the impact of our information model on the use of CAD; see also [Koo85b].

(i) the aspect of learning

The information in the creation of the design knowledge K and the implementation I, given the specification S and the language L is $H(K, I/L, S)$ and can be decomposed into:

$$H(K,I/L,S) = H(K/L,S) + H(I/K,L,S) = H(K/L,S)$$

since the term $H(I/K, L, S)$ equals zero *(FILSD)*. Hence, when the design knowledge is sufficient to derive a correct implementation I from the specification S, then the design knowledge represents the information created by the designer.

The *FILSD* has the following consequence. Since our computers as they exist today are deterministic machines, such machines cannot obtain I from S, unless K is given as some algorithm or in terms of derivation rules, or when an algorithm or rules are given by which a machine can obtain K. Our formula implies that as long as the design knowledge is such that

$$H(I/K,L,S) > 0$$

holds, we need the designer to come up with the missing knowledge. Since learning in the design process involves the creation or acquisition of knowledge that was not previously available to the designer, it is natural to assume that this learning process does not proceed without the designer making errors. This explains the occurrence of trial and error in the case of unexperienced designers. More experienced designers make less errors in this respect, since they have knowledge that enables them to derive new knowledge with a higher probability of being correct.

(ii) The man machine interface

Consider the following design system, comprising a designer (or designers) and a machine tool (or tools), like a CAD tool.

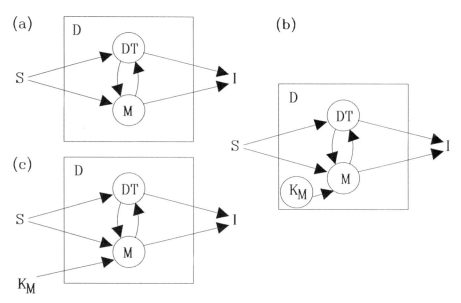

Figure 13.1: (a) A design system D with a designteam DT and computer tools M; (b) design knowledge input to the tools; (c) design knowledge input to the design system D.

Let us consider the information flows across the interface between a designer and his tools. Let DT denote the designer or design team and let M (from *Machine*) denote the tools that are used for the design task. Let $D =< DT, M >$ denote the design organization, i.e. people and machines. The information at the output of the design organization, given the specification S as input can be written as:

$$H(D/S) = H(DT/S,M) + H(M/S,DT) + T(DT : M/S).$$

The first two terms of this expression are the F_n terms of the designer and the CAD tools respectively. The third term represents the amount of information the designer has to produce in order to control the CAD tool. From the above expression one might conclude that M shows creative behaviour. From *FILSD* we know that there must be an amount of design knowledge K_M (e.g. in the form of an algorithm), such that $H(M/S, K_M) = 0$. The second term can be put to zero by explicitly stating the design knowledge K_M, which must be present within M. Then K_M can also be regarded as input to M (*Figure 13.1b*). Then the expression for $H(D/S)$ reduces to:

$$H(D/S) = H(DT,K_M/S),$$

as should be expected. Returning to the original expression for $H(D/S)$, the meaning of the third term can be deduced by considering that:

$$H(DT/S,M) = H(DT/S) - T(DT : M/S).$$

Due to the addition of the computer support, the designer's creative contribution F_n decreases by an amount which equals the information required for the coordination between man and machine. Hence the creative output of the designer is decreased by the need for controlling the machine. This requires the design of a man-machine interface which minimizes the information processing by the designer at the cognitive level at which the designer perceives the machine. It is this interface which we should optimize, e.g. by adding compute power for enhanced graphic capabilities of the CAD tool.

Since the computer takes over certain tasks, the designer can focus more directly on the design problem and hence increase the required creative output. Therefore, the net increase in the designer's noise rate ΔF_n^h can be expressed as:

$$\Delta F_n^h = \left(\begin{array}{c} \text{increase of} F_n \text{due} \\ \text{to computer support} \end{array} \right) - T(DT : M/S).$$

From this expression one may conclude that the net effect of using CAD may be negative and depends on the used man-machine interface.

13.3   Learning and knowledge representation

Design iterations are needed to support the learning process during which new de-

sign knowledge is created. Capturing design knowledge for later re-use *is essential in order to improve the effectivity of the design process.*

The capability to learn is studied extensively within the area of artificial intelligence; [MCM83] gives an introduction to this field and identifies several kinds of learning, such as: *rote learning* (learning by being programmed, or learning by memorization of given facts; no inference is needed in this case); *learning from instruction* (learning by being told; some inference is needed to integrate the new knowledge with prior knowledge); *learning by analogy* (acquiring new facts or skills by transforming and augmenting existing knowledge that bears strong similarity to the desired new concept or skill); *learning from examples* (a special case of inference, called *inductive inference* is needed in this case); *learning from observation and discovery*, which is believed to be closely related to creativity.

Although programs have been constructed that are able to learn from examples, general learning from observation and discovery (such as learning how to play chess) is an underdeveloped field. We need some fundamental breakthroughs in our understanding of learning before we can program computers to have the same capability as humans. Hence, current computers can only show a faint imitation of human learning capabilities.

Learning during design iterations will finally yield sufficient design knowledge such that *FILSD* holds and a correct implementation can be obtained. Hence, knowledge plays a key role, and it is therefore essential to try to capture such design knowledge for re-use in later designs. When asked to design, say, a CPU, the first thing the designer will do is to consider existing designs in order to re-use ideas and results from those previous designs. This re-use of design knowledge is an effective way to reduce the time needed for a new design. Yet, we very seldom document the reasoning that underlies a design with the objective to use that reasoning for a future design. Hence, it is important not only to describe the intermediate or final products of the design process, but also to describe the design process itself for the following reasons:

  (i) make the reasoning of the designer more effective;

 (ii) provide documentation support;

(iii) improve re-usability of designs and design know-how, and

(iv) enable more automation for improved design efficiency.

The notion of a *meta program* reflects this need to capture the structure and reasoning of a design process as well as the design knowledge used in it. We all have an intuitive understanding of what knowledge is. We use it in our daily lives to carry out the tasks that require some form of reasoning based on our understanding and experience. Basically there are two extreme views to the way we humans acquire knowledge: the *empirical* view and the *rationalist* view. From the empirical viewpoint, knowledge can only be derived from experience. Empirical reasoning is based on observation or experiment, not on theory. Rationalism assumes that reason based on first principles is the foundation of certainty in knowledge.

Empirical data drives our experimentation, whereas reasoning based on deduction and induction is needed to find useful conclusions (see our discussion in *Chapter 12*). Rationalism is needed to build theories from first principles (axioms), whereas empirical data are needed to test such theories against real-world phenomena. Hence, we need both views for a good balance between real-world phenomena and abstract thinking. Our reasoning from *Chapter 12* and the previous sections also indicates why the rationalistic view in knowledge is fundamentally incomplete; we need the empirical view for a very fundamental reason.

Design is an intellectual activity. Like in all intellectual activities, knowledge plays a key role. An accepted definition of knowledge is *justified true belief*. It underlies all our intelligence-based actions, but we usually do not make it explicit. It is there, but most of the time we are not aware of it. The identification and formalization of knowledge into scientific theories is what all mature disciplines of science have in common. Whether we talk about the construction of bridges, digital filters, or compilers, there is a body of knowledge which we can think of as 'mentally processed experience, information and understanding'.

By distinguishing between design objects and the processes by which these objects were obtained we are able to study their relationship in a more systematic way. We discover that the structure of a design object is intimately related to properties of mental activities that led to its creation. An understanding of the design process requires making explicit the design knowledge which plays such key role in it. We will distinguish four steps towards a formalization of design knowledge and its associated reasoning:

- *identification of design knowledge.*

This has been illustrated in *Section 8.3* by the introduction of *knowledge attributes*. They describe the knowledge used to perform a particular detailing step. In its simplest form, a knowledge attribute is described in an informal way (i.e. as a piece of text).

- *establishing a relation between knowledge attributes.*

This was done in *Section 8.3* by the introduction of the notion of *line of reasoning* and can be expressed within a meta program, which yields a

- *description of the structure of a design process.*

DPDL was introduced in *Chapter 8* and provides a syntactic structure in which pieces of informally described design knowledge can be embedded. Such a syntactic approach does not require a full formalization of all the knowledge attributes. This is in contrast with

- a formal description of design knowledge and design processes.

If we remind ourselves of the amount of knowledge associated with the simple examples in *Section 8.1*, then we get a feeling for the large amounts of knowledge to be codified if we want a full formal description of all the knowledge attributes in large designs.

Giving a formal semantics to design knowledge as well as the reasoning process is the ultimate goal, keeping in mind that also empirical design knowledge can be part of a formal structure. Full formalization is a difficult task, but we should bear in mind that partial formalizations may already enhance the effectivity of a designer considerably.

We are unable to reason about knowledge if we do not make it explicit. By **design knowledge representation** we will mean the encoding of justified true beliefs about designs or design processes. We distinguish between two major approaches in formalizing knowledge: (a) the logic approach, or (b) the object-oriented approach.

In the **logic approach**, the reasoning process as it might be carried out by a person is the main driving force behind this formalism. In first-order logic, for instance, knowledge is represented by facts and rules of deduction. Resolution and unification are the inference mechanisms by which new knowledge is inferred from existing knowledge. Logic programming is the branch of programming which uses (propositional) logic as the programming model.

In the **object-oriented approach**, not so much the potential line of reasoning, but rather the representation of things (objects) we encounter in our daily lives (or when we design artificial things such as computers) plays the dominant role. For practical situations, we will need a mixture of the two approaches.

Chapter 14

DESIGN METRICS

14.1 Design and information

Design is aimed at the transformation of a specification or high-level description into a description in terms of some real-world primitives. As such it involves the creation of information in order to remove the uncertainty about the way in which a required system can be realized.

In *Chapter 12* we have considered the occurrence of iterations from the point of view of entropy. We asked ourselves the question whether we could express the complexity of a design in order to quantify the 'size' of a design-cycle. In this chapter we will consider how to describe *structural complexity* in system descriptions in terms of both lexical content and connectivity. The relationship between the heuristic approach by Halstead [Hal79a] [Hal79b] and an approach based on information theory will be discussed. Other approaches to express structural complexity can be found in [Che78], [HK81], [McC76], [We79] and [CDS86].

A framework for the analysis of the design process, based on information theory, has been developed in *Chapters 12* and *13*; see also [Koo78] and [Koo85b]. In [Koo79] and [Koo85b] it was shown that hierarchical design considerably reduces the amount of structural information to be created by the designer, provided suitable languages are available to describe the system at different abstraction levels. It was also shown that the information content of a model is not so much a property of the model itself, but rather of the way in which the model has been obtained. Putting it another way, the structural complexity of a model is not only determined by the kind of problem for which the model is a solution, but also and above all by the organization of the (design) process which led to this model. Design is strongly related to the phenomenon of non-determinism in the sense that the outcome of the design process is unpredictable. This has been discussed in the previous chapters.

14.2   The information content of a hardware design

Our metric expresses the structural complexity of a design and is based on the
number of components in the model, the way these have been interconnected and
the level of the system description. Hence, the information content of a model is
defined relative to the previous, more abstract model from which the model was
obtained. As such, the information content is a measure of the 'distance' between
two consecutive levels. With reference to *Section 8.3*, we will focus here on the
formal model m. Let m_S denote the formal model of a system at its specification
level, and let m_I denote the implementation of m_S (*Figure 14.1*). Let L denote
the language in which m_I is expressed. In the case of an ASIC or standard cell
design, the vocabulary of L consists of the builing blocks (cells) in the ASIC or
standard cell library. We are interested to measure the information in m_I given
the specification model m_S and the language L.

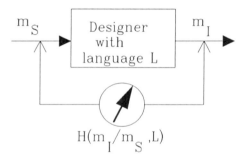

Figure 14.1: formal specification m_S is input to the designer; m_I is the output.

We can decompose the term $H(D/S)$ from *Section 12.5* also as follows:

$$
\begin{aligned}
H(D/S) &= H(K,L,I/S) \\
       &= H(L,I/S) + H(K/L,I,S) \\
       &= H(L/S) + H(I/L,S) + H(K/L,I,S)
\end{aligned}
$$

The term $H(I/L, S)$ is the information in the implementation, given the specification S and the language in which I is expressed. Writing both S and I as a
pair (m, c) we obtain:

$$
\begin{aligned}
H(I/L,S) &= H(m_I,c_I/L,m_S,c_S) \\
         &= H(c_I/L,m_S,c_S) + H(m_I/L,m_S,c_I,c_S)
\end{aligned}
$$

For the last term we find an upper bound:

$$H(m_I/L,m_S,c_I,c_S) \leq H(m_I/L,m_S)$$

The right hand side is the information in m_I given the formal model m_S and the language L. This term represents the increase in the structural information and can be measured as suggested in *Figure 14.1*. In order to do so, we will define a formal model m as a pair $(B(m), b(m))$, where $B(m)$ is the set of black boxes (language elements) used in m and $\beta(m)$ is the interconnection pattern between these black boxes (e.g. the sequential ordering of statements in a program, the nets connecting elements in a circuit diagram, etc.). Then the above expression can be decomposed into:

(14.1) $$H(m_I/m_S,L) = H(B(m_I)/m_S,L) + H(\beta(m_I)/m_S,B(m_I),L)$$

where:

$B(m_I)$ is the set of black boxes in the model m_I and
$\beta(m_I)$ is the interconnection pattern connecting these black boxes.

(In terms of Halstead's model, $B(m_I)$ corresponds to the set of elements appearing in a program string.)

The first term on the right-hand side of *(14.1)* is the information corresponding to the choice of the set of black boxes. Assuming that the vocabulary (library) of L contains η elements, and that N elements are in $B(m_I)$, then in [Fel59] it was shown that selecting these N elements from a cell library with vocabulary η can be considered as an aselect drawing with replacement of N elements from a vocabulary of η elements. This means that an element is randomly selected from the vocabulary; the result of this selection is written down after which the element is put back and a new element (possibly the same) is selected, until N elements have been selected in this way. The number of possible ways to do this is given by:

$$\binom{\eta + N - 1}{N}$$

where
η is the number of elements in the vocabulary of L, and
N is the number of elements in $B(m_I)$.

Taking the two-logarithm to obtain the information content yields:

(14.2) $$H(B(m_I)/m_S,L) = \log_2 \binom{\eta + N - 1}{N}$$

The second term on the right-hand side of *(14.1)* is the information in the interconnection pattern of m_I. In order to calculate this term, the syntax of L has to be accounted for. For our current analysis the following syntax rule will be used (the influence of syntax rules is discussed in a later section):

s1: each terminal can be connected to any other terminal except itself.

s2: In addition, it will be assumed that terminals of building blocks are unique and that connections have no direction.

Let α denote the sum of the number of terminals of m_S plus the number of terminals of all building blocks within the set $B(m_I)$. Using s1 and s2, there are

$$x = \alpha(\alpha - 1)/2$$

possible connections. As each connection may actually be available or not, this gives 2^x possible interconnection patterns. Taking the two-logarithm yields the information content of the interconnection pattern. In general, however, some of these 2^x patterns are identical, because interchanging two building blocks of the same type does not result in a different pattern. Hence, the number of possible patterns has to be divided by the number of ways in which similar black boxes can be interchanged. This number can be found by taking the product of the factorials of the number of times a building block of a certain type appears in $B(m_I)$. Taking the two-logarithm yields for the information content of the interconnection pattern:

$$\textbf{(14.3)} \qquad H(\beta(m_I)/m_S, B(m_I), L) = \alpha(\alpha - 1)/2 - \sum_{k=1}^{p} \log_2 n(k)!$$

where p is the number of different types of black boxes in $B(m_I)$ and $n(k)$ is the number of black boxes of type k.
It is assumed in formula (14.1) that m_S is a single black box. If m_S itself is made up of a number of black boxes, then the expression (14.1) is calculated for each *type* of black box, present in m_S.

As an example, consider the circuit of *Figure 14.2a*. With $\eta = 8$, $N = 32$, $\alpha = 126$, we find for the information content, relative to the language consisting of the elements {2-gate NAND, 3-gate NAND, 4-gate NAND, 2-gate OR, 3-gate OR, 4-gate OR, 2-gate AND with inverted input, 2-gate XOR}:

$$H = \log_2 \left(\frac{32 + 8 - 1}{8} \right)$$
$$+ \frac{126*125}{2} - \log_2 2! - 2\log_2 3! - 2\log_2 4! - \log_2 6! - \log_2 9!$$

The outcome of this example already suggests that the information in the interconnection pattern dominates the information content. This is due to the logarithmic dependence on the vocabulary, and the quadratic dependence of the

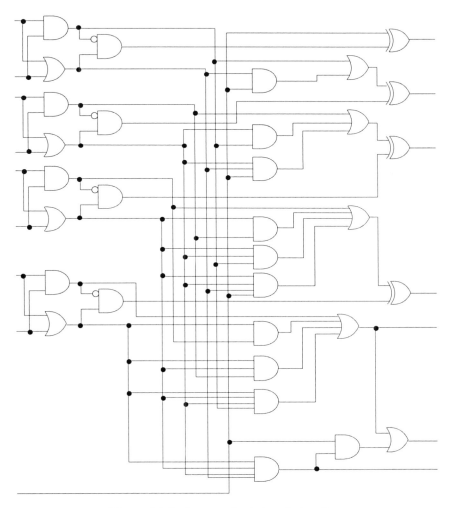

Figure 14.2: A carry-lookahead circuit.

number of interconnections. This result is also in agreement with the remark
[HK81](p.514) that *"the code length is only a weak factor in the complexity measure"* (as compared to the fan-in and fan-out component, [HK81](p513)).

14.3   Halstead's metrics

Halstead [Hal79a] introduced the concepts of *program volume, intelligence content*, etc. and he argued how these metrics relate to mental effort. Halstead's
work has been continued by several other people and some contributions can be
found in [Hal79b]. Other approaches were developed in addition to the one by
Halstead. For instance, it was felt that Halstead's metrics could not adequately
deal with the effect of control structures in programs. McCabe [McC76] suggested to use the cyclomatic complexity of flow graphs. Other metrics are based
on the number of *knots* (crossings via goto's) in Fortran programs [We79], or
the *control structural entropy* [Che78] based on the probabilities of selecting if
statements in a nested program structure. These and other methods are highly
intuitive and occupy the range of metrics in a random fashion. Henry and Kafura [HK81] present the use of information flow in programs as a more realistic
measure of program complexity.

Halstead was interested to calculate certain properties of software. He introduced expressionswith which these properties can be quantified. He bases these
expressions on the following parameters:

η_1 is the number of distinct operators appearing in a program, and

η_2 is the number of distinct operands appearing in that program.

Hence, he interprets programs in terms of operators and operands. Trivial operators are $+$, $-$ and $\times$. But also expressions like **begin...end** denote operators; in
this case these keywords denote a grouping operator. It appears that interpreting
existing programs in this way is not completely unambiguous. Thus, different
people may obtain different values for these parameters. With $\eta = \eta_1 + \eta_2$, and
N the length of the program, i.e. the total number of operators and operands
appearing in that program, Halstead defines

Program Volume: $V = N \log_2 \eta,$ *(bits)*   (14.4)

The Program Volume has the dimension of bits and represents the amount of
information in the program in terms of its vocabulary and the number of its
elements. This formula holds under the assumption that there are no constraints
regarding the exact ordering of the elements in the program sequence. In practice
there are such rules - the syntax rules - and we will consider their impact in the
next section.

Another metric is the the *Program Level* L. It reflects the 'distance' between the
actual program and its shortest possible version. It is defined as:

Program Level: $L = V^*/V$,   (14.5)

where V is the Program Volume and V^* is the Potential Volume. The latter is the volume of the shortest possible version of a program. Halstead gives an approximating expression $\hat{L}$ for L, as one may not always be able to calculate V^* directly. Using $\hat{L}$, he defines the:

Intelligence Content: $I = \hat{L} \times V$ (14.6)

Although Halstead treats V and L separately, one may conclude that I is an approximation of V^* if one replaces $\hat{L}$ by L in (3): $I = L \times V = (V^*/V) \times V = V^*$. Hence the distinction between V^* and I made by Halstead is not a conceptual one, but is due to an approximation in one of his heuristic expressions (that of L).

In summary: the Program Volume V measures the amount of information in a program, the Program Level is a measure of program 'shortness', whereas the Intelligence Content is a measure of the amount of information in the shortest version of a program.

14.4   The information content of a software program

Information is a measure of uncertainty; when applied to a message channel it shows how much uncertainty is taken away upon the arrival of a certain message. It is important to realize that information is always defined relative to a fixed set of "things", like a set of possible messages; upon receipt of a particular message a certain amount of information is received. The information model explicitly states this principle by calculating the number of possible designs ("messages") which can be built ("sent") using a given number of black boxes (the symbols of the message). Then the design of a particular model corresponds to the receipt of a particular message from a set of possible messages (the possible designs).

Halstead's measure is based on sequential interconnection patterns, since he is concerned with the sequential nature of programs. In our information model, a distinction has been made between the building blocks of a formal model and its interconnection pattern. This also opens up the possibility of making the influence of syntax rules more explicit [Koo79].

Syntax rules have a large impact on the information content. Halstead considered sequential constructs (programs) and the syntax rules which he incorporated in calculating the Program Volume were of the type $<$ operator $><$ operand $>$. The actual syntax rules of the programming language put more severe constraints on the constructs which can be made. This means that the building blocks in a program do not have a uniform frequency distribution in general. The effect of syntax rules can best be illustrated with a small example.

Suppose a language L_a has a vocabulary of two elements as shown in *Figure 14.3a*. Hence, $\eta = 2$ and from the figure it follows that $N = 10$, $n(1) = 5$ and $n(2) = 5$ (the model contains five instances of each type of element). Then, according to formulas *(14.1)*, *(14.2)*, and *(14.3)* the model m_1 as shown in *Figure 14.3b* has the following information content:

$$H(m_I/m_S, L_a) = \log_2 \begin{pmatrix} 2 & + & 10 & - & 1 \\ & & 10 & & \end{pmatrix} + \tfrac{22(22-1)}{2} - \log_2 5!5!$$

$$= log_2 11 + 231 - 2(3 + log_2 3 - log_2 5)$$

Now suppose L_a contains a the following syntax rules:

(r1): *an element of type A can only be connected to elements of type B and vice versa.*

(r2): *connections are unidirectional.*

Then the information in m_I, using these syntax rules, is:

$H(m_I/m_S, L_a, r1, r2) = 1 + \log_2 10,$

which illustrates the large influence of syntax rules.

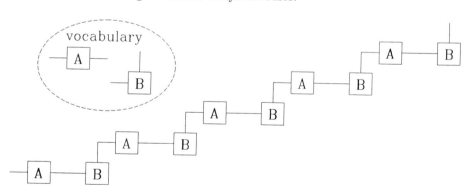

Figure 14.3: (a) The vocabulary of the language La; (b) a model expressed in this language.

Next, we will analyze the sequential ordering as used by Halstead. Suppose the language in which the model is expressed has a vocabulary of three elements $\{A, B, C\}$. Hence: $\eta = 3$. Suppose, a model is built containing three elements, i.e. $N = 3$. Then according to (1), but without taking the logarithm, the Program Volume (the number of different sequences) is $3 \times 3 \times 3 = 27$. For the sake of comparison, the resulting sequences have been arranged in a certain way:

Sequences	Interconnection pattern
AAA	$3!/3! = 1$
BBB	
CCC	
AAB ABA BAA	$3!/1!2! = 3$
AAC ACA CAA	
ABB BAB BBA	
ACC CAC CCA	
BBC BCB CBB	
BCC CBC CCB	
ABC ACB BAC BCA CAB CBA	$3!/1!1!1! = 6$

The total number of sequences is 27, the number given by the Program Volume
formula (without taking the logarithm). The number of sequences in the first
column is 10. This is the same as predicted by the formula *(14.2)* (without
taking the logarithm); this formula calculates the number of different sequences,
irrespective of the order in which the elements appear in the sequences. The
ordering is calculated separately using the interconnection pattern approach.
Formula *(14.3)* cannot be used as it will be assumed that the interconnection
pattern is purely sequential (i.e. we will incorporate a syntax rule prescribing
this ordering).

In that case three elements can be arranged in 3! different ways, provided the
elements are distinct. If two are equal, this number should be divided by 2!
(this is the correction factor introduced in *(14.3)*). In general, the number of
permutations $N!$ of N elements should be divided by the product of the factorials
of the number of times different elements occur in the sequence. There are
three patterns consisting of identical elements (per pattern there is $3!/3! = 1$
possibility). There are six sets of patterns containing two identical elements;
within each set there are $3!/1!2! = 3$ different patterns with the same pair of
identical elements. And there is one set of patterns containing different elements;
within this set there are $3!/!1!1!1 = 6$ different patterns.

In conclusion, the information content of a program P(in terms of its structural complexity) relative to a language L can be calculated using the following
formula:

$$(14.7) \qquad H(P/L) = \log_2 \binom{\eta + N - 1}{N} + \log_2 N! - \sum_i \log_2 n(i)!$$

where η is the number of elements in the vocabulary of L and $n(i)$ is the number
of elements of type i within the sequence. Consider the following descriptions of
a NAND and a NOR circuit, expressed in LISP:

(DEFINEQ NAND(X Y)   DEFINEQ NOR(X Y)

$$\text{(COND(X (NOT Y))} \qquad\qquad \text{(COND(NOT X)(NOT Y))}$$
$$\text{(T T)))} \qquad\qquad\qquad \text{(T F)))}$$

LISP is considered here to have a vocabulary which consists of the following elements:

1. the 98 functions as listed in appendix 2 of [WH81];

2. the operand separator *()*;

3. the space separator;

4. the type *variable*, and

5. the type *constant*.

This gives a value for η of 102. In the description of NAND, the type *variable* as well as the type *constant* appear in different forms. For example, X and Y are both of the type *variable*, but they are certainly distinct. In order to incorporate the fact that variables and constants can be distinct, one should interpret the factor $n(i)$ of *(14.7)* as the number of similar elements, distinct from the other elements in the sequence. Applying *(14.7)* to NAND gives:

$$\log_2 \binom{102\ +19\ -1}{19} + \log_2 19! - \log_2 6! - \log_2 2! - \log_2 2!.$$

And for NOR:

$$\log_2 \binom{102\ +24\ -1}{24} + \log_2 24! - \log_2 8! - 3\log_2 2!$$

14.5 Relating the Information Model with Halstead's metrics

Referring to our design process model as presented in *Chapter 8*, we will assume that the version of a program as used in the Potential Volume is in fact the shortest possible program and therefore corresponds to our initial model m_0. Hence, the Potential Volume represents the amount of information in m_0 relative to some initial language L_0:

(14.8)   Potential Volume $V^* = H(m_0/L_0)$

Likewise, Halstead's Program Volume measures the information in some implementation m_i of the program at level i in terms of the vocabulary of a language L_i at that level. Hence:

(14.9a)   Program Volume $V = H(m_i/L_i)$

With a reasonable approximation this can be written as:

(14.9b) $V = H(m_i/m_0, L_i)$

Hence, Halstead does not relate the information at a certain level to the previous model of the sequence (as has been done in the Information Model), but to the initial model of the sequence. Combining *(14.4)*, *(14.8)*, and *(14.9)* we find:

(14.9)   Program Level $L = H(m_0/L_0)/H(m_i/L_i)$

A disadvantage of this approach is that successive models at different levels are not related in this information measure. Our Information Model explicitly links models at successive levels by stating the information content of a model (implementation) in terms of the previous model (specification) and the design language which is used in the corresponding design cycle. This way of considering matters seems more natural as one considers the input and output of the designer. Halstead does not state the principle of hierarchical design explicitly, although he states that a program should not contain identical substrings of length η, i.e. there should not be any redundancy. Dealing with redundancy at the highest possible abstraction level is one of the characteristics of hierarchical design.

About the Intelligence Content Halstead states: *"It would seem that a funda-mental measure of how much is said in a program should properly be called its information content, but the meaning of that term has been preempted by Shan-non's Information Theory to refer only to the volume of a message, ignoring its level."*

However, as the Intelligence Content is conceptually equivalent to the Potential Volume, and hence can be expressed in terms of *(14.8)*, the level of the imple-mentation is also ignored by Halstead as it only refers to the initial model, which is supposed not to alter.

What Halstead would like to call the information content (and which he actually calls the intelligence content), is not suitable for expressing the information of a model at a certain level. In fact, it does not measure the information content.

Henry and Kafura [HK81] use a term $\alpha_{in} \times \alpha_{out}$ (where α_{in} is the number of local flows into a procedure plus the number of data structures from which this procedure retrieves information; α_{out} is the number of outflows plus the number of data structures updated by the procedure); $\alpha_{in} \times \alpha_{out}$ represents the total possible number of combinations of an input source to an output destination. This is in agreement with our expression for the number of interconnection pat-terns with the distinction that rules $s1$ and $s2$ are not used by them. The term $\alpha_{in} \times \alpha_{out}$ represents the *internal* complexity of a procedure. They square this term because *"the weighting of the fan-in and fan-out component is based on the belief that the complexity is more than linear in terms of the connections which a procedure has to its environment".* What they actually do is to apply the expression for the interconnection pattern to include the number of ways a procedure can be connected to its environment; hence, the quadratic dependence in: $(\alpha_{in} \times \alpha_{out})^2$. The complexity of a procedure is then found by multiplying this term with N (the lenght of a procedure in terms of number of lines of text). Taking the two-logarithm we find for the information content in the Henry/Kafura approach:

$$\log_2 N + 2\log_2(\alpha_{in}) + \log_2(\alpha_{out}).$$

The dependence on the interconnection pattern is logarithmic rather than linear as in *(14.2)*.

14.6   Development time estimation

A pragmatic approach to software engineering economics can be found in [Boe81]. Halstead derived an expression for predicting the time necessary to develop a program of a certain volume. This relation has been criticized by Coulter [Cou83]; he argues that the time measure given by Halstead which is based on the famous Stroud number [Str66], is disputable from the cognitive point of view. However, the idea to estimate design time by taking the information content of a design and deviding this by a factor with the dimension of *bits/time-unit* is an interesting one. In terms of our information model, we would get:

design time :: $H(I/L, S)/T$

where T would be the required productivity factor. Halstead uses the already mentioned Stroud number, the value of which is assumed to be in the order of 5-10 bits/second. We have to bear in mind that a design of an optimal software program may yield less code (and, hence, less structural complexity or information content), but requires a longer time to produce. Optimization is an activity which contributes strongly to increasing development times. However, for straightforward designs (for example, an ASIC design which would not require very difficult timing optimization) the above approach looks very promising.

Part IV

APPENDIX

Appendix A

The behaviour of a multiplier

In the following expansion, within sum expressions only terms with priority equal to one (indicated by **) are expanded further; zero priority terms are not expanded as they represent behaviours which will never be selected in the sum behaviour. First, we will write down the equations for the subsystems, parameterized with the right values.

P1| CNT(i) = (0) t1!: (P2| CNT(i)) + (1) t4!: (P1| CNT(i))
P2| CNT(i) = (0) t2!: (P1| CNT(i)) + (1) t3!: (P2| CNT(i+1))
PX(3) = t1!: PX(2)
PX(2) = t1!: PX(1)
PX(1) = t1!: NIL
PY(2) = PY(2)| P3(0)
PY(2)| P3(0) = t3!: (PY(1)| P3(1))
PY(1)| P3(1) = t3!: (PY(0)| P3(2)) + t4!: (PY(2)| P3(0))
PY(0)| P3(2) = t4!: (PY(1)| P3(1))

We are now able to calculate the folowing expression:

P1| PX(3)| PY(2) = P1| CNT(0)| PX(3)| PY(2)| P3(0)
P1| CNT(0)| PX(3)|PY(2)| P3(0) =
          (0) τ: P2| CNT(0)| PX(2)| PY(2)| P3(0)
P2| CNT(0)| PX(2)| PY(2)| P3(0) =
          t2! : P1| CNT(0)| PX(2)| PY(2)| P3(0)0
          + (1) τ : P2| CNT(1)| PX(2)| PY(1)| P3(1))**
P2| CNT(1)| PX(2)| PY(1)| P3(1) =
          t2! : P1| CNT(1)| PX(2)| PY(1)| P3(1)0
          +(1) τ : P2| CNT(2)| PX(2)| PY(0)| P3(2)**
P2| CNT(2)| PX(2)| PY(0)| P3(2) =
          (0) t2!: P1| CNT(2)| PX(2)| PY(0)| P3(2)
P1| CNT(2)| PX(2)| PY(0)| P3(2) =
          (0) τ : P2| CNT(2)| PX(1)| PY(0)| P3(2)

$$+(1)\ \tau : \text{P1}|\ \text{CNT(2)}|\ \text{PX(2)}|\ \text{PY(1)}|\ \text{P3(1)}^{**}$$

P1| CNT(2)| PX(2)| PY(1)| P3(1) =
$$\tau : \text{P2}|\ \text{CNT(2)}|\ \text{PX(1)}|\ \text{PY(1)}|\ \text{P3(1)0}$$
$$+(1)\ \tau : \text{P1}|\ \text{CNT(2)}|\ \text{PX(2)}|\ \text{PY(2)}|\ \text{P3(0)}^{**}$$

P1| CNT(2)| PX(2)| PY(2)| P3(0) =
$$(0)\ \tau: \text{P2}|\ \text{CNT(2)}|\ \text{PX(1)}|\ \text{PY(2)}|\ \text{P3(0)}$$

P2| CNT(2)| PX(1)| PY(2)| P3(0) =
$$\text{t2!} : \text{P1}|\ \text{CNT(2)}|\ \text{PX(1)}|\ \text{PY(2)}|\ \text{P3(0)0}$$
$$+(1)\ \tau : \text{P2}|\ \text{CNT(3)}|\ \text{PX(1)}|\ \text{PY(1)}|\ \text{P3(1)}^{**}$$

P2| CNT(3)| PX(1)| PY(1)| P3(1) =
$$\text{t2!} : \text{P1}|\ \text{CNT(3)}|\ \text{PX(1)}|\ \text{PY(1)}|\ \text{P3(1)0}$$
$$+(1)\ \tau : \text{P2}|\ \text{CNT(4)}|\ \text{PX(1)}|\ \text{PY(0)}|\ \text{P3(2)}^{**}$$

P2| CNT(4)| PX(1)| PY(0)| P3(2) =
$$(0)\ \text{t2!:}\ \text{P1}|\ \text{CNT(4)}|\ \text{PX(1)}|\ \text{PY(0)}|\ \text{P3(2)}$$

P1| CNT(4)| PX(1)| PY(0)| P3(2) =
$$\tau : \text{P2}|\ \text{CNT(4)}|\ \text{NIL}|\ \text{PY(0)}|\ \text{P3(2)0}$$
$$+(1)\ \tau : \text{P1}|\ \text{CNT(4)}|\ \text{PX(1)}|\ \text{PY(1)}|\ \text{P3(1)}^{**}$$

P1| CNT(4)| PX(1)| PY(1)| P3(1) =
$$\tau : \text{P2}|\ \text{CNT(4)}|\ \text{NIL}|\ \text{PY(1)}|\ \text{P3(1)0}$$
$$+(1)\ \tau : \text{P1}|\ \text{CNT(4)}|\ \text{PX(1)}|\ \text{PY(2)}|\ \text{P3(0)}^{**}$$

P1| CNT(4)| PX(1)| PY(2)| P3(0) =
$$(0)\ \tau: \text{P2}|\ \text{CNT(4)}|\ \text{NIL}|\ \text{PY(2)}|\ \text{P3(0)}$$

P2| CNT(4)| NIL| PY(2)| P3(0) =
$$\text{P2}|\ \text{CNT(4)}|\ \text{PY(2)}|\ \text{P3(0)}$$

P2| CNT(4)| PY(2)| P3(0) =
$$\text{t2!} : \text{P1}|\ \text{CNT(4)}|\ \text{PY(2)}|\ \text{P3(0)0}$$
$$+(1)\ \tau : \text{P2}|\ \text{CNT(5)}|\ \text{PY(1)}|\ \text{P3(1)}^{**}$$

P2| CNT(5)| PY(1)| P3(1) =
$$\text{t2!} : \text{P1}|\ \text{CNT(5)}|\ \text{PY(1)}|\ \text{P3(1)0}$$
$$+(1)\ \tau : \text{P2}|\ \text{CNT(6)}|\ \text{PY(0)}|\ \text{P3(2)}^{**}$$

P2| CNT(6)| PY(0)| P3(2) =
$$(0)\ \text{t2!:}\ \text{P1}|\ \text{CNT(6)}|\ \text{PY(0)}|\ \text{P3(2)}$$

P1| CNT(6)| PY(0)| P3(2) =
$$(1)\ \tau: \text{P1}|\ \text{CNT(6)}|\ \text{PY(1)}|\ \text{P3(1)}$$

P1| CNT(6)| PY(1)| P3(1) =
$$(1)\ \tau: \text{P1}|\ \text{CNT(6)}|\ \text{PY(2)}|\ \text{P3(0)}$$

P1| CNT(6)| PY(2)| P3(0) = NIL

(multiplier halts in a state, where PY has 2 tokens and PX is empty and the result of the multiplication is in CNT)

Substitution results in the following behaviour of the multiplier:

P1 | PX(3) | PY(2) =
$$(0)\tau{:}(1)\tau{:}(1)\tau{:}\text{t2!:}(1)\tau{:}(1)\tau{:}(0)\tau{:}(1)\tau{:}(0){:}\ \tau{:}(0)\text{t2!:}(1)\tau{:}(1)\tau{:}$$

system engineering a

Order Date: 01/14/92
Number : 516679-47971
Year : 1991

507059B Copy bs 316810

Total Charge : 62.48

Rush: _____

Title: The design of communicating systems
          pproach / by C.J. Koomen.

Author: Koomen, C.J. (Cees-Jan), 1947-

Edition: Volume: Cor
Comments:

Publisher:ACADEMIC PRESS

ISBN: 0-7923-9203-5 (acid-free

Copies: 1   Estimated Cost: 30.00

Supplier: Total Information Inc.
          844 Dewey Avenue
          Rochester, NY 14613-0202

Total Info In Stock? no

~~~~~~

$(0)\tau:(1)\tau:(1)\tau:(0)$t2!:$(1)\tau:(1)\tau:$ P1| CNT(6)| PY(2)| P3(0)1

Removing the priorities and applying the first τ-law yields:

P1 | PX(3) | PY(2) = τ: t2!: t2!: t2!: P1| CNT(6)| PY(2)| P3(0).

As we are not interested in the explicit knowledge of when t2 fires (we are only interested to know the result of the firing of the multiplier as a whole), we hide this firing by replacing t2 by τ. Applying the 1st τ-law again yields:

P1 | PX(3) | PY(2) = τ: P1| CNT(6)| PY(2)| P3(0)

Appendix B

Expansion for the network equations

Application of the combination algorithm leads to the following set of equations (*ax* means *axioma*):

G0G0	$= $ G.in1? con : G1G0(con)	(1)
G1G0(con)	$= $ G.in1? ster : G0G0(con,ster)	(2)
	$+ \tau$: G1G0(con*)	
G0G0(con,ster)	$= $ G0G0	(ax) (3)
G1G0(con*)	$= $ G.in1? ster : G0G0(con*,ster)	(4)
	$+ $ G.out2! con : G1G3	
G0G0(con*,ster)	$= $ G.out2! con : G0G3(ster)	(5)
	$+ \tau$: G0G0(con*,ster*)	
G1G3	$= $ G.in1? ster : G0G3(ster)	(6)
	$+ $ G.in2? ptr : G1G4(ptr)	
	$+ $ G.in2? ntr : G1G0(ntr)	
G0G3(ster)	$= \tau$: G0G3(ster*)	(7)
	$+ $ G.in2? ptr : G0G4(ster,ptr)	
	$+ $ G.in2? ntr : G0G0(ster,ntr)	
G0G3(ster*)	$= $ G.out2! ster :G0G0	(8)
	$+ $ G.in2? ptr : G0G4(ster*,ptr)	
	$+ $ G.in2? ntr : G0G0(ster*,ntr)	
G0G0(con*,ster*)	$= $ G0G0	(ax) (9)
G1G4(ptr)	$= $ G.in1? ster : G0G4(ster,ptr)	(10)
	$+ $ G.in2? dter : G1G0(ptr,dter)	
	$+ \tau$: G1G4(ptr*)	
G1G4(ptr*)	$= $ in1? ster : G0G4(ster,ptr*)	(11)
	$+ $ G.in2? dter : G1G0(ptr*,dter)	
	$+ $ G.out1! ptr : G2G4	
G1G0(ntr)	$= $ G.in1? ster : G0G0(ster,ntr)	(12)

	$+\ \tau\ :\ \text{G1G0(ntr}^*)$	
G1G0(ntr*)	$=\ \text{G.in1? ster : G0G0(ster,ntr}^*)$	(13)
	$+\ \text{G.out1! ntr : G5G0}$	
G0G4(ster,ptr)	$=\ \tau\ :\ \text{G0G4(ster}^*,\text{ptr})$	(14)
	$+\ \tau\ :\ \text{G0G4(ster,ptr}^*)$	
	$+\ \text{G.in2? dter : G0G0(ster,ptr,dter)}$	
G0G4(ster*,ptr)	$=\ \text{G.out2! ster : G0G6(ptr)}$	(15)
	$+\ \tau\ :\ \text{G0G4(ster}^*,\text{ptr}^*)$	
	$+\ \text{G.in2? dter : G0G0(ster}^*,\text{ptr,dter)}$	
G0G4(ster,ptr*)	$=\ \tau\ :\ \text{G0G4(ster}^*,\text{ptr}^*)$	(16)
	$+\ \text{G.in2? dter : G0G0(ster,ptr}^*,\text{dter)}$	
G0G4(ster*,ptr*)	$=\ \text{G.out2! ster : G0G6(ptr}^*)$	(17)
	$+\ \text{G.in2? dter : G0G0(ster}^*,\text{ptr}^*,\text{dter)}$	
G0G0(ster,ntr)	$=\ \tau\ :\ \text{G0G0(ster}^*,\text{ntr})$	(18)
	$+\ \tau\ :\ \text{G0G0(ster,ntr}^*)$	
G0G0(ster*,ntr)	$=\ \text{G0G0}$	(ax) (19)
G0G0(ster,ntr*)	$=\ \text{G0G0}$	(ax) (20)
G1G0(ptr,dter)	$=\ \text{G1G0(ntr)}$	(ax) (21)
G2G4	$=\ \text{G.in1? ster : G0G4(ster)}$	(22)
	$+\ \text{G.in2? dter : G2G0(dter)}$	
G1G0(ptr*,dter)	$=\ \text{G.in1? ster : G0G0(ster,ptr}^*,\text{dter)}$	(23)
	$+\ \text{G.out1! ptr : G2G0(dter)}$	
	$+\ \tau\ :\ \text{G1G0(ptr}^*,\text{dter}^*)$	
G1G0(ptr*,dter*)	$=\ \text{G1G0(ntr}^*)$	(ax) (24)
G0G0(ster,ptr,dter)	$=\ \text{G0G0(ster,ntr)}$	(ax) (25)
G0G0(ster*,ptr,dter)	$=\ \text{G0G0(ster}^*,\text{ntr})$	(ax) (26)
G0G0(ster,ptr*,dter)	$=\ \tau\ :\ \text{G0G0(ster}^*,\text{ptr}^*,\text{dter)}$	(27)
	$+\ \tau\ :\ \text{G0G0(ster,ptr}^*,\text{dter}^*)$	
G0G0(ster*,ptr*,dter)	$=\ \tau\ :\ \text{G0G0(ster}^*,\text{ptr}^*,\text{dter}^*)$	(28)
G0G0(ster,ptr*,dter*)	$=\ \text{G0G0(ster,ntr}^*)$	(ax) (29)
G0G0(ster*,ptr*,dter*)	$=\ \text{G0G0(ster}^*,\text{ntr}^*)$	(ax) (30)
G0G0(ster*,ntr*)	$=\ \text{G0G0}$	(ax) (31)
G0G6(ptr)	$=\ \text{G.in2? dter : G0G0(ptr,dter)}$	(32)
	$+\ \tau\ :\ \text{G0G6(ptr}^*)$	
G0G6(ptr*)	$=\ \text{G.in2? dter : G0G0(ptr}^*,\text{dter)}$	(33)
G0G0(ptr,dter)	$=\ \text{G0G0(ntr)}$	(ax) (34)
G0G0(ntr)	$=\ \text{G0G0}$	(ax) (35)
G0G0(ptr*,dter)	$=\ \tau\ :\ \text{G0G0(ptr}^*,\text{dter}^*)$	(36)
G0G0(ptr*,dter*)	$=\ \text{G0G0}$	(ax) (37)
G5G0	$=\ \text{G.in1? ster : G0G0(ster)}$	(38)
G0G0(ster)	$=\ \tau\ :\ \text{G0G0(ster}^*)$	(39)
G0G0(ster*)	$=\ \text{G0G0}$	(ax) (40)
G0G4(ster)	$=\ \tau\ :\ \text{G0G4(ster}^*)$	(41)

	+ G.in2? dter : G0G0(ster,dter)	
G0G4(ster*)	= G.out2! ster : G0G6	(42)
	+ G.in2? dter : G0G0(ster*,dter)	
G2G0(dter)	= G.in1? ster : G0G0(ster,dter)	(43)
	+ τ : G2G0(dter*)	
G2G0(dter*)	= G.in1? ster : G0G0(ster,dter*)	(44)
	+ G.out1! dter : G5G0	
G0G0(ster,dter)	= τ : G0G0(ster*,dter)	(45)
	+ τ : G0G0(ster,dter*)	
G0G0(ster*,dter)	= G0G0	(ax) (46)
G0G0(ster,dter*)	= G0G0	(ax) (47)
G0G6	= G.in2? dter: G0G0(dter)	(48)
G0G0(dter)	= G0G0	(ax) (49)

Appendix C

Reduction of the equations

The reduction is done starting at the last equation and then working backwards towards the first equation. The following renaming of identifiers needs hindsight, but it will simplify our equations here; in practice this renaming can only be done after the reduction):

G0G0	->	G0	G0G4(ster*,ptr*)	->	G7
G1G0(con*)	->	G1	G0G4(ster*)	->	G7'
G0G0(con*,ster)	->	G2	G2G4	->	G8
G1G3	->	G3	G2G0(dter*)	->	G9
G0G3(ster*)	->	G4	G1G0(ptr*,dter)	->	G10
G1G4(ptr*)	->	G5	G5G0	->	G11
G1G0(ntr*)	->	G6	G0G6	->	G12

Then perform the following reductions (from left to right, and top to bottom):

(49) -> (48)	(47) -> (45,44)	(46) -> (45,42)
Sum4 (42)	(45) -> (43,41)	τ-1 (43,41)
τ-4 (43); i.e. G2G0(dter) = τ: G9		
(43) -> (22,23)	τ-1 (22,23)	τ-(41)
(41) -> (22)	τ-1 (22)	(40) -> (39)
(39) -> (38)	τ-1 (38)	(37) -> (36)
(36) -> (33)	τ-1 (33)	(35) -> (34)
(34) -> (32)	(33) = (48); Therefore: G0G6(ptr*) = G12	
τ-4 (32)	(32) -> (15)	τ-1 (15)
(31) -> (30)	(30) -> (28)	(20) -> (29,18,13)
(29) -> (27)	(28) -> (27,17)	τ-1 (27,17)
Sum4 (27)	(27) -> (23,16)	τ-1 (23,16)
(19) -> (26,18,8)	(26) -> (15)	Sum4 (18)
(18) -> (7,12)	τ-1 (7,12)	(18) -> (25)

(25) -> (14)	τ-1 (14)	(24) -> (23)
τ-4 (12)	(12) -> (21,6)	τ-1 (6)
(21) -> (10)	τ-1 (10)	We see that G7'= G7
τ-4 (16)	(16) -> (14,11)	τ-1 (14,11)
τ-4 (15)	(15) -> (14,8)	τ-1 (14,8)
Sum4 and τ-4 (14)	(14) -> (10,7)	τ-1 (10,7)
(9) -> (5)	τ-4 (7)	(7) -> (5,6)
τ-1 (5,6)	(3) -> (2)	

Substituting G5 (11) and G10 (23) -> (10) we see that:

G1G4(ptr) = τ:(G.in1? ster: G7
                    + G.out1! ptr: G8
                    + G.in2? dter:
                            (G.in1? ster: G0
                             + G.out1! ptr: G9
                             + τ: G6))
        + G.in2? dter: G6
        + G.in1? ster: G7

The last term can be removed (τ-4); the remaining expression is of the form:

G1G4(ptr) = τ: (A + g: (B + τ: G6)) + g: G6,

where:

  A = G.in1? ster: G7 + G.out1! ptr: G8
  B = G.in1? ster: G0 + G.out1! ptr: G9
  g = G.in2? dter

The right-hand side can be replaced by a simpler expression as follows:

τ: (A + g: (B + τ: G6)) + g: G6
τ: (A + g: (B + τ: G6) + g: G6) + g: G6   (τ-3)
τ: (A + g: (B + τ: G6) + g: G6)   (τ-4)
τ: (A + g: (B + τ: G6)).   (τ-3)

Hence G1G4(ptr) = τ: G5.

(10) -> (6) τ-1 (6)   (5) -> (4)
(4) -> (2);

The remaining expression (2) is of the form:

G1G0(con) = g: G0 + τ: (g: (A + τ: G0) + B),

where g = G.in1?ster; A = G.out2!con: G4 and B = G.out2!con: G3. This can
be replaced by:

$$\text{G1G0(con)} = \text{g: G0} + \tau\text{: (g: G0} + \text{g: (A} + \tau\text{: G0)} + \text{B)} \qquad (\tau\text{-}3)$$
$$= \tau\text{: (g: G0} + \text{g: (A} + \tau\text{: G0)} + \text{B)} \qquad (\tau\text{-}4)$$
$$= \tau\text{: (g: (A} + \tau\text{: G0)} + \text{B)} \qquad (\tau\text{-}3)$$

Hence, G1G0(con) = τ: G1.

(2) -> (1) $\qquad\qquad$ τ-1 (1).

Appendix D

Network equations

After the reduction as carried out in *Appendix C*, one obtains the following equations, which specify the behaviour of the network G towards a source node and the corresponding destination node.

G0 = in1? con : (G1 + τ : G6)
G1 = G.in1? ster: G2   + G.out2! con: G3
G2 = G.out2! con: G4   + τ: G0
G3 = G.in1? ster: G4   + G.in2? ptr: G5   + G.in2? ntr: G6
G4 = G.out2! ster: G0   + G.in2? ptr: G7   + G.in2? ntr: G0
G5 = G.in1? ster: G7   + G.out1! ptr: G8 + G.in2? dter: G10
G6 = G.in1? ster: G0   + G.out1! ntr: G11
G7 = G.out2! ster: G12 + G.in2? dter: G0
G8 = G.in1? ster: G7   + G.in2? dter: G9
G9 = G.in1? ster: G0   + G.out1! dter: G11
G10 = G.in1? ster: G0   + G.out1! ptr: G9 + τ: G6
G11 = G.in1? ster: G0
G12 = G.in2? dter: G0

Bibliography

[Bae86]   J.C.M. Baeten. *Procesalgebra: een formalisme voor parallel, communicerende processen.* Kluwer programmatuurkunde, 1986.

[Bea84]   R.K. Brayton and et al. *Logic Minimization Algorithms for VLSI synthesis.* Kluwer Academic Press, Boston, 1984.

[BK83]   J.A. Bergstra and J.W. Klop. A Process Algebra for the Operational Semantics of Static Data Flow Networks. Preprint IW 222/83, february 1983.

[Boe81]   B.W. Boehm. *Software Engineering Economics.* Prentice Hall, 1981.

[Boe88]   B.W. Boehm. A spiral model of software Development and Enhancement. *IEEE computers*, pages 61–72, may 1988.

[Bra80]   W. Brauer, editor. *Net theory and applications. Proceedings of the Advanced Course on General Net Theory of Processes and Systems.* Springer Verlag, Berlin Heidelberg New-York, 1980.

[Bri62]   L. Brillouin. *Science and information theory*, chapter 20. Academic Press, New York, second edition, 1962.

[Bro75]   F.P. Brooks. *The Mythical Man Mouth.* Addison-Wesley, Reading , Mass., 1975.

[CC90]   P.M. Churchland and P. Smith Churchland. Could a Machine Think. *Scientific American*, 262(1):26–31, january 1990.

[CCI88a]   CCITT. Functional Specification and Description Language SDL, Recommendations Z.100 and Annexes A-F, 1988.

[CCI88b]   CCITT. Specifications of Signalling System No.7, Recommendation Q.700-Q.716, Q.721-Q.766, Q.771-Q.795, 1988.

[CDS86]   S.D. Conte, H.E. Dunsmore, and V.Y. Shen. *Software Engineering Metrics and Models.* Benjamin/Cummings Publishing Compagny, California, 1986.

[CFPS86]   J.P. Crutchfield, J.D. Farmer, N.H. Packard, and R.S. Shaw. Chaos.
            Scientific American, 255(6):38–49, December 1986.

[Che78]   E.T. Chen. Program Complexity and Programmer Productivity.
            IEEE Trans. Software Eng., SE-4:187–194, May 1978.

[Con76]   R.C. Conant. Laws of Information which govern Systems. *IEEE
            Trans.*, SMC-6(4):240–255, 1976.

[Cou83]   N.S. Coulter. Software Science and Cognitive Psychology. *IEEE
            Trans. Software Eng.*, SE-9:166–171, March 1983.

[CP88]   P. Camurati and P. Prinetto. Formal Verification of Hardware Cor-
            rectness: Introduction and Survey of Current Research. *Computer*,
            21(7):8–20, July 1988.

[CY90]   P. Coad and E. Yourdan. *Object oriented analysis*. Yourdan press,
            Prentice Hall, 1990.

[Deo74]   N. Deo. *Graph Theory with Applications to Engineering and Com-
            puter Science*. PrenticeHall, Englewood Cliffs, 1974.

[DF84]   E.W. Dijkstra and W.H.J. Feijen. *Een Methode van Programmeren
            ('A Programming Method')*. Academic Services, The Hague, 1984.

[dG88]   P.J. de Graaff. Private Communications. Eindhoven University of
            Technology, private communication, 1988.

[DH87]   E. Dubois and J. Hagelstein. Reasoning on Formal Requirements: A
            Lift Control system. In *Proceedings of the 4th International Work-
            shop on Software Specification and Design*, pages 161–168, Mon-
            terey, California, april 1987.

[DHL+86]   E. Dubois, J. Hagelstein, E. Lahou, F. Ponsaert, A. Rifaut, and
            F. Williams. The ERAE model: A case study. In T.W. Olle, H.G.
            Sol, and A.A. Verrijn-Stuart, editors, *Information Systems Design
            Methodlogies: Improving the practice*, pages 87–105. North Holland,
            1986.

[Dij76]   E.W. Dijkstra. *A Discipline of Programming*. Prentice Hall, Engle-
            wood Cliffs, 1976.

[dMCG+90]   H. de Man, F. Catthoor, G. Goossens, J. Vanhoof, J. van Meerber-
            gen, S. Note, and J. Huisken. Architecture-Driven Synthesis tech-
            niques for VLSI Implementation of DSP Algorithms. In *Proceedings
            of the IEEE*, volume 78 (2), pages 319–335, february 1990.

[dMRC86]   H. de Man, J. Rabaey, and L. Claesen. Cathedral-ii: A silicon com-
           piler for digital signal processing. *Computer*, pages 13–25, december
           1986.

[Dow86]   M. Dowson, editor. *Iteration in the Software Process: Proceed-
           ings of the 3rd Software Process Workshop*, Breckenridge, Colorado,
           November 1986. IEEE Computer Society Press.

[Fel59]   W. Feller. An introduction to information theory and its applica-
           tions, 1959.

[Gel87]   P. Gelli. Evaluation and Comparison of Three Specification Lan-
           guages: SDL, Lotos and Estelle. In R. Saracco and P.A.J. Tilanus,
           editors, *SDL-87 - State of the Art and Future Trends*. Elsevier Sci-
           ence Publishers, 1987.

[GJ79]   M.R. Garey and D.S. Johnson. *Computers and Intractability; A
           guide to the theory of NP-completeness*. Freeman, 1979.

[Gor83]   M.J.C. Gordon. LCF-LSM. Technical Report 41, University of
           Camebridge, Computer Laboratory, 1983.

[Gri84]   D. Gries. *The Science of Computer Programming*. Springer Verlag,
           New York, Heidelberg, Berlin, 1984.

[Hal79a]   M.H. Halstead. *Elements of software science*, volume I of *The Com-
           puter Science Library*. North Holland, New York, 1979.

[Hal79b]   Commemorative issue in honor of Dr. Maurice H. Halstead, March
           1979.

[Hen88]   M. Hennessy. *Algebraic Theory of Processes*. MIT-Press, 1988.

[HitV90]   R.J. Huis in 't Veld. The Role of Languages in the Design-trajectory.
           In Don Fay and Lorenzo Mezzalira, editors, *EUROMICRO '90*,
           pages 177–183, Amsterdam, Holland, august 1990. North Holland.

[HK81]   S. Henry and D. Kafura. Software Structure Metrics Based on Infor-
           mation Flow. *IEEE Trans. Software Eng.*, SE-7, September 1981.

[Hoa85]   C.A.R. Hoare. *Communicating Sequential Processes*. Prentice Hall,
           1985.

[Hog89]   D. Hogrefe. *Estelle, Lotos und SDL: Standard Spezifikation-
           ssprachen f r Verteilte Systeme*. Springer Verlag, Berlin, 1989.

[Hop84]   J.E. Hopcroft. Turing Machines. *Scientific American*, 250(5):70–80,
           May 1984.

[HP87]   D.J. Hatley and I Pirbhai. *Strategies for Real-time System Specification*. Dorset House Publishing Co., 1987.

[Jac86]   M.A. Jackson. *System Development*. Prentice Hall, 1986.

[Jon86]   C.B. Jones. *Systematic Software Development Using VDM*. Prentice Hall, 1986.

[JV80]   M. Jantzen and R. Valk. Formal properties of place/transition nets. In *Net theory and applications. Proceedings of the Advanced Course on General Net Theory of Processes and Systems*, pages 165–212. Springer Verlag, Berlin Heidelberg New-York, 1980.

[Kal86]   A. Kaldewaij. *A formalism for concurrent processes*. PhD thesis, Eindhoven University of Technology, 1986.

[Koo78]   C.J. Koomen. Information laws for system design. In *Proc. IEEE Int. Conf. on Cybernetics and Society*, Tokyo, November 1978.

[Koo79]   C.J. Koomen. Reducing model complexity in system design. In *Proc. IEEE Int. Conf. on Cybern. and Society*, pages 830–833, Denver, October 1979.

[Koo84]   C.J. Koomen. Thinking about Software Design: A Meta Activity. In Colin Potts, editor, *Proceedings of Software Process Workshop*, pages 19–24, Egham, Surrey, UK, february 1984. IEEE Computer Society Press.

[Koo85a]   C.J. Koomen. Algebraic Specification and Verification of Communication Protocols. *Science of Computer Programming*, 5:1–36, january 1985. North-Holland.

[Koo85b]   C.J. Koomen. The Entropy of Design: a Study on the Meaning of Creativity. *IEEE Trans. on Systems, Man and Cybernetics*, SMC-15(1):16–30, January 1985.

[Koo86]   C.J. Koomen. Iterations, Learning and the Detailing Step Paradigm. In M Dowson, editor, *Proceedings of the 3rd International Software Process Workshop*, pages 36–39, Breckenridge Colorado, USA, november 1986. Computer Society Press of the IEEE.

[Koo88]   C.J. Koomen. Limits to the Mechanization of the Detailing Step Paradigm. In C. Tully, editor, *Proceedings of the 4rd International Software Process Workshop*, pages 97–102, Moretonhampstead, Devon , UK, may 1988. ACM Press.

[Law90]   H.W. Lawson. Philosophies for Engineering Computer-Based Systems. *IEEE Computer*, pages 52–63, December 1990.

[LB85]   M.M. Lehman and L.A. Belady. *Program Evolution*. Academic Press, 1985.

[LSU90]   R. Lipsett, C.F. Schaefer, and C. Ussery. *VHDL: hardware description and design*. Kluwer Academic Press, 2 edition, 1990.

[LZ77]   B. Liskov and S. Zilles. An introduction to formal specification of data abstractions. In Yeh R.T., editor, *Current trends in programming methodology*, volume I, chapter 1, pages 1–32. Prentice-Hall, Inc., Englewood Cliffs, 1977.

[MC80]   C. Mead and L. Conway. *Introduction to VLSI Systems*. Addison-Wesley, 1980.

[McC76]   T. McCabe. A Complexity Measure. *IEEE Trans. Software Eng.*, SE-2:308–320, December 1976.

[MCM83]   R.S. Michalski, J.G. Carbonell, and T.M. Mitchell. *Machine Learning*. Tioga Publishing Co, Palo Alto, California, 1983.

[Mil78]   R. Milner. Synthesis of communicating behaviour. In J Winkowski, editor, *Proceedings 7th symposium on mathematical foundations of computer science*. Lecture notes of computer science, Springer Verlag, Berlin Heidelberg, New-York, 1978.

[Mil80]   R. Milner. *A Calculus of Communicating Systems*. Lecture Notes in Computer Science 92. Springer Verlag, Berlin Heidelberg New-York, 1980.

[Mil87]   R.E. Miller. The Construction of Self-synchronizing Finite State Protocols. *Distributed Computing*, 2:104–112, 1987.

[Mil89]   R. Milner. *Communication and Concurrency*. Prentice Hall, 1989.

[MM90]   ed. M. Moriconi. Formal Methods in Software Development. In *Proceedings of the ACM SIGSOFT workshop*, Napa, California, May 1990.

[Mrd90]   S. Mrdalj. Biblio of Object-Oriented System Development. *ACM SIGSOFT Software Engineering Notes*, 15(5):60–63, october 1990.

[NN58]   E. Nagel and J.R. Newman. *G dels Proof*. New York University Press, New York, 1958.

[OP90]   F. Orava and J. Parrow. Algebraic Descriptions of Mobile Networks: An Example. In L. Logrippo, R.L. Probert, and H. Ural, editors, *The 10th international IFIP WG 6.1 Symposium on protocol specification, testing, and verification*, Ottawa, Ontario, Canada, June 1990. North-Holland.

[Org84]   International Standards Organization. ISO/IS7498 open systems interconnection–basic reference model, 1984.

[Par72]   D.l. Parnas. On criteria to be used in decomposing systems into modules. *Communications of the ACM*, 15(12):1053–1058, 1972.

[Pet81]   J.L. Peterson. *Petri net theory and the modeling of systems.* Prentice-Hall, Englewood Cliffs, 1981.

[Pra86]   Praxis. The ELLA language reference manual, 1986.

[PS75]   D.L. Parnas and D.P. Siewiorek. Use of the concept of transparency in the design of hierarchically structured systems. *Comm. of the ACM*, 18(7):401–408, 1975.

[Rem85]   M. Rem. Concurrent Computations and VLSI Circuits. *Control Flow and Data Flow: Concepts of Distributed Programming*, pages 399–437, 1985.

[ROL90]   S. Rugaber, S.B. Ornburn, and R.J. LeBlanc. Recognizing Design Decisions in Programs. *IEEE Software*, pages 46–54, January 1990.

[Sea90]   J.R. Searle. Is the Brain's Mind a Computer Program? *Scientific American*, 262(1):20–25, january 1990.

[Shi82]   M.W. Shields. Concurrency, Correctness, Proof and Undecidability in SDL-like Systems. Internal Report CSR-119-82, University of Edinburgh, Department of Computer Science, june 1982.

[Sim68]   H.A. Simon. *The Sciences of the Artificial.* MIT Press, Cambridge Ma. and London UK, 8 edition, 1968.

[SP90]   A.P. Sage and J.D. Palmer. *Software Systems Engineering.* Wiley, New York, 1990.

[SPT87]   R. Saracco and eds. P.A.J. Tilanus. *SDL87 - State of the Art and Future Trends.* Elsevier Science Publishers, 1987.

[Str66]   J.M. Stroud. The fine structure of psychological time. *Annals of the New York Academy of Sciences*, pages 623–631, 1966.

[Tan88]   A.S. Tanenbaum. *Computer Networks.* Prentice Hall, 1988.

[Tre87]   L. Trevillyan. Overview of logic synthesis. In *Proceedings IEEE Design Automation Conference*, 1987.

[We79]   M.R. Woodward and et.al. A Measure of Control Flow Complexity in Program Text. *IEEE Trans. Software Eng.*, SE-5, January 1979.

[WH81]   P.H. Winston and B.K.P. Horn. *LISP.* Addison-Wesley, 1981.

[Win90]   J.M. Wing. A Specifier's Introduction to Formal Methods. *IEEE Computer*, pages 8–24, September 1990.

[WM85]   P.T. Ward and S.J. Mellor. *Structured Development for Real-Time Systems.* Prentice Hall, 1985.

[Zav82]   P. Zave. An operational approach to requirements specification for embedded systems. *IEEE Trans. Software Eng.*, SE-8(3):250–269, May 1982.

Index